Site Engineering for
LANDSCAPE ARCHITECTS WORKBOOK

Site Engineering for
LANDSCAPE ARCHITECTS WORKBOOK

SECOND EDITION

To Accompany Site Engineering for Landscape Architects, Sixth Edition

Jake Woland

WILEY

John Wiley & Sons, Inc.

Cover Illustration: Courtesy of Jake Woland
Cover Design: David Riedy

This book is printed on acid-free paper. ⊗
Copyright © 2013 by John Wiley & Sons, Inc. All rights reserved.

Published by John Wiley & Sons, Inc., Hoboken, New Jersey.
Published simultaneously in Canada.

For general information about our other products and services, please contact our Customer Care Department within the United States at (800) 762-2974, outside the United States at (317) 572-3993 or fax (317) 572-4002.

Wiley publishes in a variety of print and electronic formats and by print-on-demand. Some material included with standard print versions of this book may not be included in e-books or in print-on-demand. If this book refers to media such as a CD or DVD that is not included in the version you purchased, you may download this material at http://booksupport.wiley.com. For more information about Wiley products, visit www.wiley.com.

Library of Congress Cataloging-in-Publication Data:

978-1-118-09085-5

Printed in the United States of America

10 9 8 7 6 5 4 3 2 1

Contents

Introduction

This workbook has been introduced to complement the 6th edition of *Site Engineering for Landscape Architects*. It has been designed as a study tool to reinforce concepts from the textbook. The questions presented in the workbook can be used in the classroom, as well as by individuals as a self-study tool, with the ultimate goal of helping prepare individuals for taking the LARE or other licensing exams.

The workbook is organized with a chapter of questions and a separate chapter of answers that correspond to the respective chapter in the textbook. As necessary, tables and graphics from the textbook required to solve questions posed in the workbook have been reprinted in the appropriate chapters. The questions in each chapter are generally of four types:

1. Observations
2. Short-answer questions
3. Long-answer questions
4. Graphic questions

In some chapters, all four types of questions may be presented, whereas in others, only one or two types may be presented. This depends on the material being covered. A detailed explanation of the different types of questions follows.

OBSERVATIONS

Landscape architecture is a profession that requires lifelong learning. The observation questions in the workbook are meant to call your attention to elements of designing with landforms that pose particular challenges that you will encounter throughout your career. These observations will serve as a baseline of inspiration for dealing with the different types of design problems you will face in the coming years. It is recommended that you keep a journal that traces these observations with both narrative and images. Develop this documentation in whatever form you are comfortable with.

Photo documentation of these observations is critical in order to have the information available as you progress in your career. It is highly recommended that you spend time not only photographing your observations but also geolocating them along with the pertinent narrative information using software such as Google Earth™ for later use. A Facebook page (www.facebook.com/SiteEngineeringForLandscapeArchitects) has been created to provide a consolidated location for students to upload geolocated photos with a brief description in the different categories of observation provided throughout the workbook. The author will curate these images into a growing compendium of examples of excellence in site engineering. Your participation is encouraged to make this a robust resource. Students submitting excellent photographs may be solicited for inclusion of the photographs in future versions of the textbook.

Answers for observation questions are not provided in the Answers section of the workbook.

SHORT-ANSWER QUESTIONS

The short-answer questions may take several forms. Fill-in-the-blank, multiple-choice, and mathematical word questions will be combined throughout, depending on the material being emphasized. These questions are meant to reinforce the material presented in each chapter and offer a way to quickly study the basic information in the chapter.

LONG-ANSWER QUESTIONS

The long-answer questions are used to focus attention on multifaceted concepts and more complex issues of design. They require a more in-depth understanding of the basic information in the chapter.

GRAPHIC QUESTIONS

The graphic questions offer opportunities to apply the knowledge gained from the chapter. They are designed to increase in complexity within each chapter. In most cases, the answers presented in the Answers section of the workbook are but one of various possible solutions to each question.

In summary, the workbook is designed to help beginning designers build confidence in using the concepts of site engineering while also allowing practitioners to sharpen their skills in preparation for the licensing exam. It will also help broaden the understanding of the material through observation and application of knowledge, which are both critical to successfully applying these skills to real projects.

QUESTIONS

CHAPTER 1

Questions

1.1

Explore as many of the following types of landscapes as possible, both early in your study of this material and after you have developed a greater understanding of the material. As you experience these places, think about how the use of landform in the design affects your experience. Does it provide a sequence of experience? Does it evoke a certain feeling or emotion? How does the landform interact with plantings and other designed elements to create the overall composition? Are there things about the landform that you would change to improve your experience of the place?

Take photographs of the places you explore and upload the photos to the Site Engineering for Landscape Architects Facebook page (https://www.facebook.com/SiteEngineeringForLandscape Architects). Please provide geolocation information and a brief description of your observation with any uploaded photos.

Types of places to explore:
- Residences of different types and scales
- State and national parks
- University or college campuses
- Corporate campuses
- Athletic fields
- Civic spaces/places—city hall, post office, library, courthouse
- Places of worship

1.2

Search your local area and identify roads of the following types that you will revisit throughout your study of site engineering. As you take the time to explore and identify these roads, observe how the road and its adjacent conditions interact. Can you tell whether the road has been fit into the landscape, leaving its surroundings largely intact, or whether the landscape has been altered to accommodate the road and adjacent development?

Take photographs of the streets, roads, and highways and upload the photos to the Site Engineering for Landscape Architects Facebook page.

Road types to assess:

- Residential street in a new development.
- Residential street in an older part of town.
- A stretch of highway with a diversity of different landscapes adjacent to it. This could include urban, suburban, and rural residential development; farmland; and industrial and natural landscapes.
- A winding rural road.
- A road in a local, state, or national park.

1.3

Find a construction site near where you live that will be relatively easy to visit and observe over time. (Obtain official permission as necessary to enter each location for observation.) You will be visiting this site as you learn more about the various topics involved with site engineering. The more complex the construction, the better example it will be to use in this series of observations. To show the progress of the construction, take photographs of the site at least weekly.

Take photographs of the construction site and upload the photos to the Site Engineering for Landscape Architects Facebook page.

1.4

Find a local habitat restoration to visit. (Obtain official permission as necessary to enter each location for observation.) Research in advance a description of the landscape development of the restoration. What role did landform play in restoring this habitat? As you visit the site, can you tell that there has been construction on the site? Does the restoration have a clear boundary, or does the site blend well with the surrounding landscape?

Take photographs of the habitat restoration and upload the photos to the Site Engineering for Landscape Architects Facebook page.

1.5

Sketch, photograph, or create a collage of one of your favorite places, designed or not. To accompany your imagery, write a narrative of the place that answers the following questions:

- How is the character of this place affected by the landforms of which it is composed?
- How do the landforms interact with the plant life to create the place?
- How would you change the landform to improve this place?
- If it is a designed landscape, to what type of character described in the textbook does it most closely correspond?
- Do the different spatial considerations mentioned in the reading play a part in your experience of the place?

- Are there particular environmental functions that appear to be a part of the composition of the place?

1.6

Model in the medium with which you are most comfortable (either digital or physical) a landform or series of landforms that you feel conveys one or more of the emotions below. If you are creating a physical model, make it at least 12" × 12" in size.

Take photographs of screen shots of your model and upload them to the Site Engineering for Landscape Architects Facebook page.

- anger
- joy
- sadness
- excitement
- surprise
- fear
- hope

CHAPTER 2

Questions

2.1

Walk through your campus or town. Look at how the different buildings meet grade and how the architecture relates to the topography within which it is set. Are there relationships that are typical for particular types of buildings?

Take photographs of these relationships and upload the photos to the Site Engineering for Landscape Architects Facebook page.

2.2

Drive the roads you identified in Chapter 1.
- Look at how the street grid relates to the surrounding topography.
- Also observe how the buildings adjacent to the roadway are situated with respect to it. Are they

depressed below it, built up to sit above it, or on an even level with it?

2.3

Visit the construction site you chose to observe in Chapter 1. Look for signs of erosion and erosion control structures on the site. Are there places where erosion seems to be occurring without a structure to control it, or are there structures that appear to have failed? Is any existing vegetation being protected on the site?

Also observe the texture of the soil on the site. Can you see sand grains in the soil or even stones and pebbles? Does sediment make the runoff cloudy? This suggests a large component of fine-grained silt and/or clay particles.

2.4

Take a walk on a rainy day. Look at how the rain flows across the landscape.

Are the drainage features hard (concrete and metal drain inlets with the water piped away underground) or soft (vegetated swales) systems or a combination of the two? Does water pool in places where it should not? Are there places where this seems to happen typically? Why do you think this is the case?

2.5

Look up your local municipal code. (Most can be found online.) Look in the table of contents for sections that apply to site engineering. This might include topics related to criteria for fire access, construction in floodplains, setbacks from property lines for construction, or storm water management. From your review of these documents, compile a list of constraints to landform design outlined by the code.

2.6

Look up the latest Americans with Disabilities Act regulations at www.ada.gov. In particular, review requirements for accessible routes, ramps, stairs, and handrails. What constraints do these regulations put on the design of landforms?

Now explore your local area and look at how different buildings and landscapes have been designed to comply with these requirements. Find and document both situations in which the requirements are well integrated into the design

and others in which that is not the case. What reasons can you think of for these different responses?

Take photographs of these access features and upload the photos to the Site Engineering for Landscape Architects Facebook page.

2.7

Increase in impervious surface in construction results in (increased/decreased) fluctuation of water levels in streams, ponds, and wetlands and (increased/decreased) potential for flooding.

2.8

In order to best protect existing plant material, where should grade change be limited to with respect to the plant material?

2.9

List three types of soils that are typically avoided or removed when identified on a project site.

2.10

What are two topographic factors that influence a site's potential for erosion?

2.11

Name three design criteria that constrain site engineering that are typically regulated by a political authority.

2.12

When connecting a newly developed storm drainage system to an existing storm drainage system, what is the best starting point for establishing elevations in the design of the new system?

2.13

What type of foundation provides the greatest amount of grading flexibility and potentially the least amount of grading impact?

2.14

Match the uses below with their desired ranges of slope.

1. Terraces and sitting areas   a. 0.5 to 1.5%
2. Planted banks   b. 1 to 2%
3. Playfields   c. 2 to 3%
4. Public streets   d. 1 to 5%
5. Parking areas   e. 1 to 8%
6. Game courts   f. up to 50%

2.15

What is the phrase used to describe designing so that storm water is drained away from a structure to avoid structural and moisture problems?

2.16

What is the legal extent to which grades can be changed on a particular site?

CHAPTER 3

Questions

3.1

Walk through your campus or town. Look for places where contour lines represent themselves in the landscape and document them. This might include coursing of masonry on buildings, the tops of level walls, and the edges of fountains or natural bodies of water.

Take photographs of these contour lines and upload the photos to the Site Engineering for Landscape Architects Facebook page.

3.2

Find a large sandbox or sand volleyball court (and get permission to use them if they do not belong to you). If neither is available, work with your classmates to construct a 3' × 3' × 10" sandbox for your studio. Use the sandbox to explore creating abstract forms that incorporate the contour signatures described in Chapter 3, including ridges; valleys; summits; depressions; and uniform, concave, and convex slopes. After creating an abstract landform that includes at least three different contour signatures, draw contour lines that show how these different signatures intersect.

3.3

Take a walk on a rainy day. Knowing that water flows perpendicularly to contours, imagine how the contours would need to be drawn on elements in the landscape to allow water to move the way it is. These observations should include both impervious surfaces such as roads and sidewalks as well as soft landscaped areas.

3.4

Find a place that has a path and a wall or curb that are sloping (and whose owners will not mind your drawing in chalk on them). Draw contour lines as you understand them to work across various hardscape elements. After drawing the contour lines, check them using a carpenter's level, or use a level app on your smartphone.

3.5

Now that you have taken some time to examine contours in the field, look through the pictures that you have taken to document your observations so far. Find two that show a distinct landform and more than 4 ft of grade change. One should include naturalistic curves and the other should be architectonic. At least one should have a grade change device in the frame of the picture. Now take copies of these pictures and trace contours as you imagine them to work over top. If it is easier to take chalk into the field to start thinking about this, do so.

3.6

What is the contour interval of the portion of the USGS map shown in Figure 3.1, shown in feet of elevation?

3.7

A (concave/convex/uniform) slope is one in which the slope gets progressively steeper moving from higher to lower elevations.

Figure 3.1. Contour map excerpt.

3.8

What is the mistake in the contours drawn in Figure 3.2?

3.9

How many contour lines are required to indicate a three-dimensional form and direction of slope?

3.10

The steepest slope is (parallel/perpendicular) to a contour line.

3.11

Construct a section of the topography shown in Figure 3.3 at the cut line indicated.

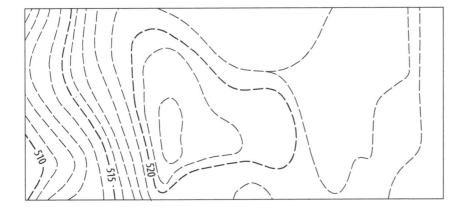

Figure 3.2. Contour plan

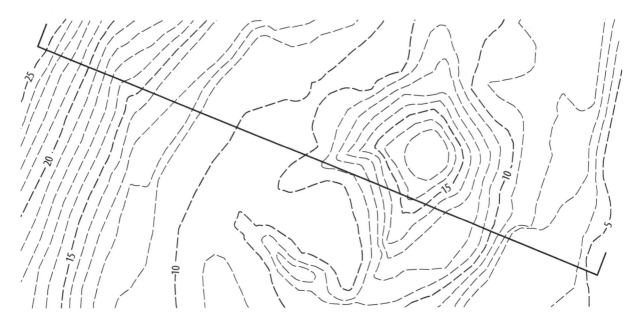

Figure 3.3. Contour plan

3.12

On the topography presented in Figure 3.4, identify the areas that show the following contour signatures:

- summit
- depression
- even slope
- concave slope
- convex slope

3.13

Build a model of the topography shown in Figure 3.4. To do this, make two copies at 400 percent. This should make the image 10" × 10". Attach each copy to a 10" × 10" piece of cardboard. Next, cut all the even contour lines out of one piece of cardboard and all the odd contour lines out of the other. Now stack them, interlacing odds and evens. Glue them together in order to finish your model.

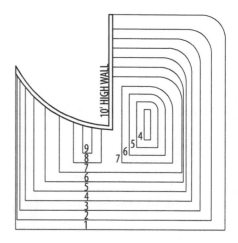

Figure 3.4. Contour plan

Take a look at the completed model. Does it change your ideas about the location of any of the contour signatures?

3.14

Identify the contour signatures for the areas called out on the map in Figure 3.5.

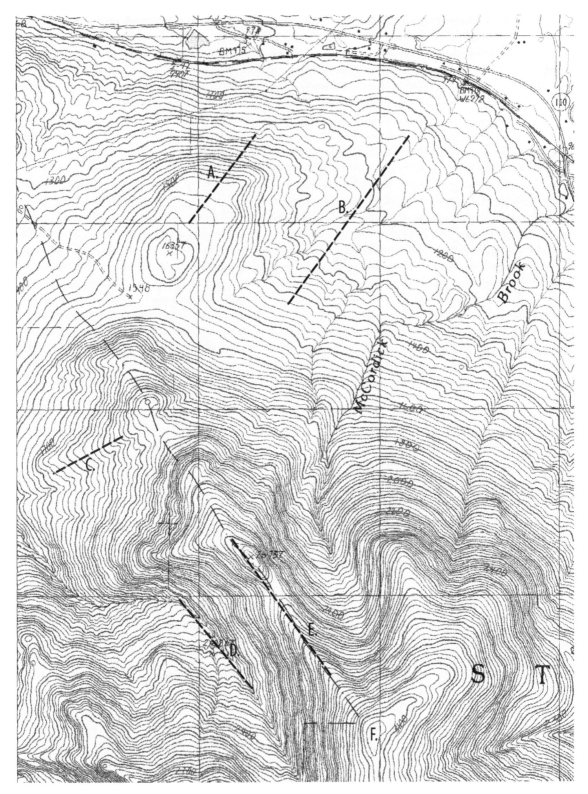

Figure 3.5. Contour map excerpt

CHAPTER 4

Questions

4.1

In Figure 4.1, draw the 1-ft contours that result from interpolating between the elevations provided.

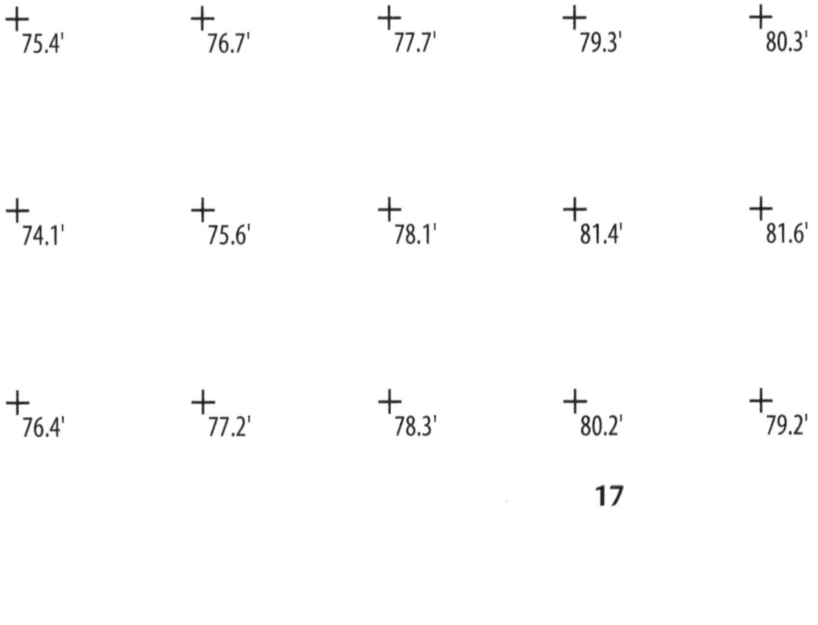

$+$
75.4'
$\quad$ $+$
76.7'
$\quad$ $+$
77.7'
$\quad$ $+$
79.3'
$\quad$ $+$
80.3'

$+$
74.1'
$\quad$ $+$
75.6'
$\quad$ $+$
78.1'
$\quad$ $+$
81.4'
$\quad$ $+$
81.6'

$+$
76.4'
$\quad$ $+$
77.2'
$\quad$ $+$
78.3'
$\quad$ $+$
80.2'
$\quad$ $+$
79.2'

Figure 4.1. Spot elevations for interpolation

4.2

For the information provided in Figure 4.2, determine elevations for points A, B, and C.

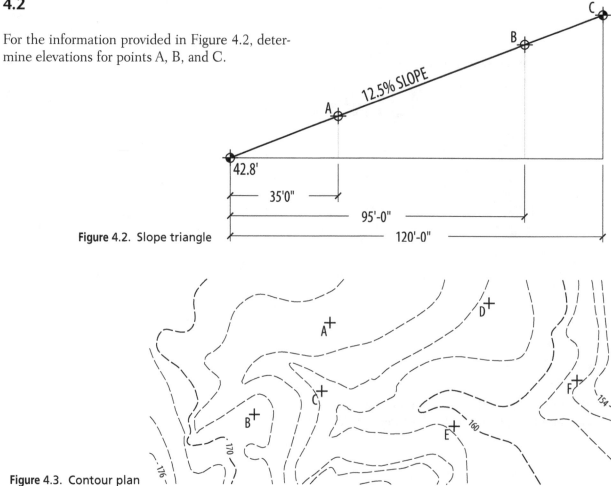

Figure 4.2. Slope triangle

Figure 4.3. Contour plan

4.3

Interpolate the elevations at the points shown Figure 4.3.

following points if the drawing is at a scale of 1" = 30'-0": AC, AD, BC, CE, CF, DF, EF.

4.4

Having identified the elevations of the various points in Figure 4.3, find the slope between the

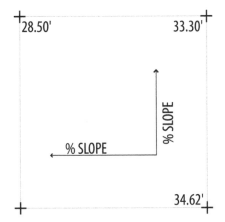

Figure 4.4. Slope calculation diagram

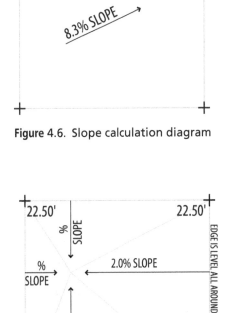

Figure 4.6. Slope calculation diagram

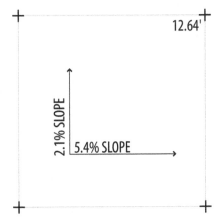

Figure 4.5. Slope calculation diagram

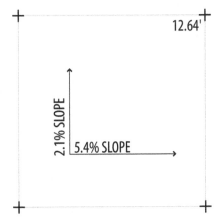

Figure 4.7. Slope calculation diagram

4.5

Calculate the missing information from the information provided in each of the following diagrams (Figures 4.4, 4.5, 4.6, and 4.7). Arrows indicate direction of descending slope. Each square is drawn at a scale of 1" = 20'-0". Round elevations off to the nearest hundredths and slopes off to tenths.

4.6

Convert the following ratios to percentage slopes: 50:1, 10:1, 8:1, 3:1.

4.7

Convert the following percentage slopes to degrees and minutes: 1.5 percent, 3.5 percent, 18 percent, 35 percent.

4.8

Map the following slope categories onto the topography shown in Figure 4.8:

- $0 \leqslant 5\%$,
- $> 5 \leqslant 20\%$,
- $>20\%$

The topography is drawn at a scale of 1" = 50'-0".

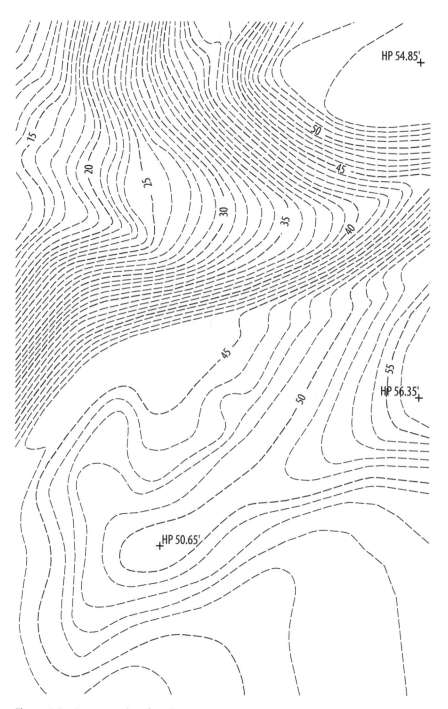

Figure 4.8. Contour plan for slope analysis

CHAPTER 5

Questions

5.1

The slope perpendicular to traffic on a sidewalk is called the _____. It is typically graded at _____ percent.

5.2

Using the information provided, calculate the cross slope represented in Figure 5.1, which is drawn at a scale of 1" = 10'-0".

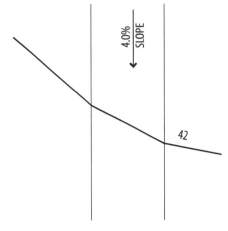

Figure 5.1. Cross-slope calculation diagram

5.3

Using the information provided in Figure 5.2, draw in the contour that will represent the cross slope shown. Provide a smooth transition when tying back into the existing contour. The figure is drawn at a scale of 1" = 10'-0".

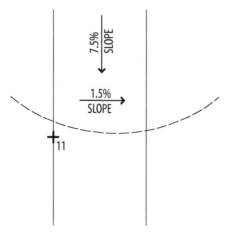

Figure 5.2. Cross-slope contour diagram

5.4

Design the centerline of a path to connect between two existing path segments at points A and B in Figure 5.3 with a maximum gradient of 5 percent. The figure is drawn at a scale of 1" = 30'-0", and the contour interval is 1'-0".

5.5

What are the two purposes of a road crown?

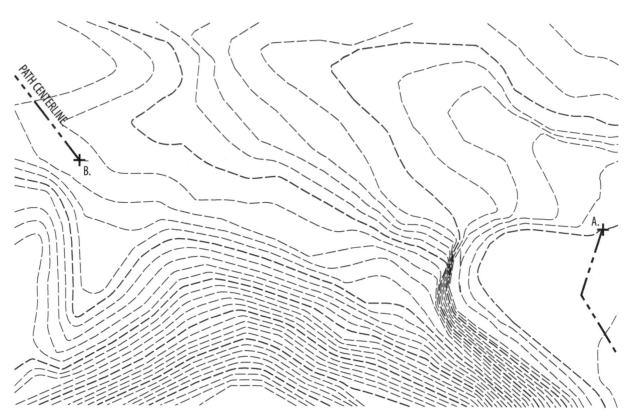

Figure 5.3. Contour plan for path design

5.6

Given the information provided in the following figures (Figures 5.4, 5.5, 5.6, and 5.7), draw in the 1-ft contours. Each figure is drawn at a scale of 1" = 20'-0".

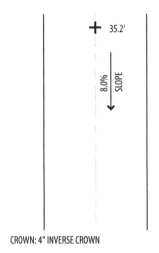

CROWN: 4" INVERSE CROWN

Figure 5.4. Road grading plan

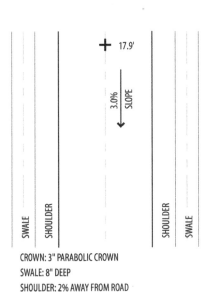

CROWN: 3" PARABOLIC CROWN
SWALE: 8" DEEP
SHOULDER: 2% AWAY FROM ROAD

Figure 5.5. Road grading plan

5.7

Given the information provided in Figure 5.8, draw in the 1-ft contours. Assume that slopes are even between the noted high point and the other two spot elevations. The plan is drawn at a scale

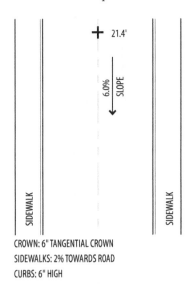

CROWN: 6" TANGENTIAL CROWN
SIDEWALKS: 2% TOWARDS ROAD
CURBS: 6" HIGH

Figure 5.6. Road grading plan

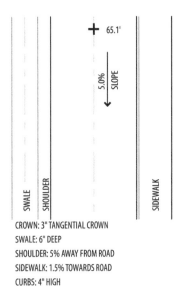

CROWN: 3" TANGENTIAL CROWN
SWALE: 6" DEEP
SHOULDER: 5% AWAY FROM ROAD
SIDEWALK: 1.5% TOWARDS ROAD
CURBS: 4" HIGH

Figure 5.7. Road grading plan

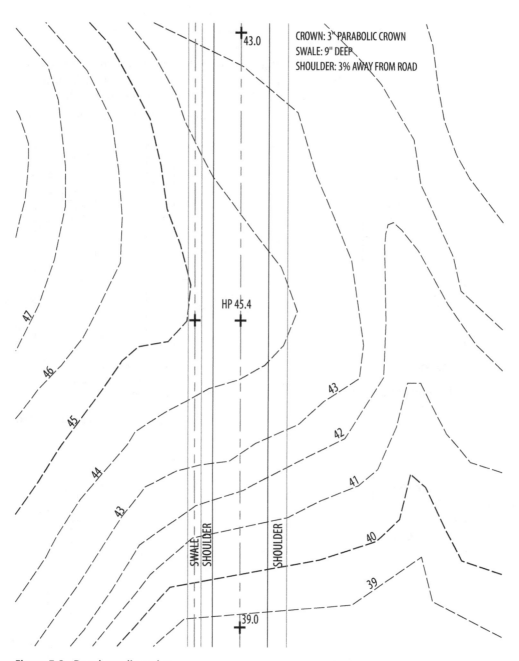

CROWN: 3" PARABOLIC CROWN
SWALE: 9" DEEP
SHOULDER: 3% AWAY FROM ROAD

Figure 5.8. Road grading plan

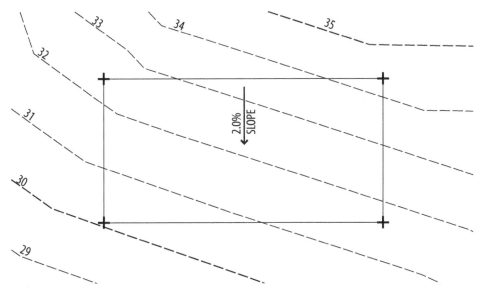

Figure 5.9. Terrace grading plan

of 1" = 40'-0". Provide smooth transitions when tying back into the existing contours.

5.8

Providing drainage away from a structure is referred to as providing _____ drainage.

5.9

Grade the terrace located on the plan in Figure 5.9 so that it is completely on fill and has positive drainage provided on at least three sides. No slope shall be greater than 3:1. The scale of the plan is 1" = 20'-0".

5.10

Grade the terrace located on the plan in Figure 5.10 so that it is partially on cut and partially on fill and has positive drainage provided on all sides. No slope shall be greater than 4:1. The scale of the plan is 1" = 20'-0".

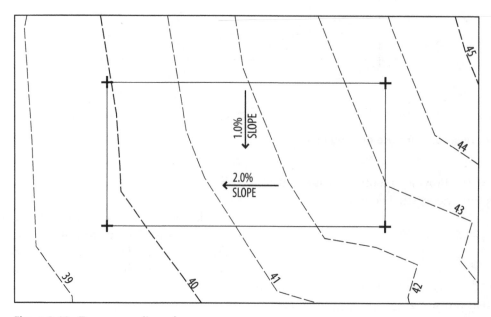

Figure 5.10. Terrace grading plan

CHAPTER 6
Questions

6.1

Grade the entry plaza to the building located on the plan in Figure 6.1 using the information provided. Use spot elevations, accompanied by intended slopes to define the grading between the building entrance and the sidewalk spot elevations, instead of using contours. Use the following criteria in design:

- No slope shall be greater than 5.0 percent, except on the ADA ramp, which can be 8.3 percent.
- No slope shall be less than 2.0 percent.
- Stair risers will be between 4 in. and 7 in. high and evenly spaced.
- Stair treads will be between 12 in. and 15 in. in depth.
- All stair runs shall have one extra tread width at the top and bottom of the stair to accommodate railings.

- Stairs are allowed a 0.1-ft change in elevation across the face of the riser where they meet a sloping grade.
- Extend the length of the planter walls toward the street as needed to provide separation between sets of stairs.
- Document the number and size of treads and risers for each stair.
- The scale of the plan is 1" = 10'-0."

6.2

In the plan shown in Figure 6.2, provide access to the four townhomes from the sidewalk using the following criteria:

- Use combinations of stairs and ramps to provide access, without disturbing the grades at the sidewalk.

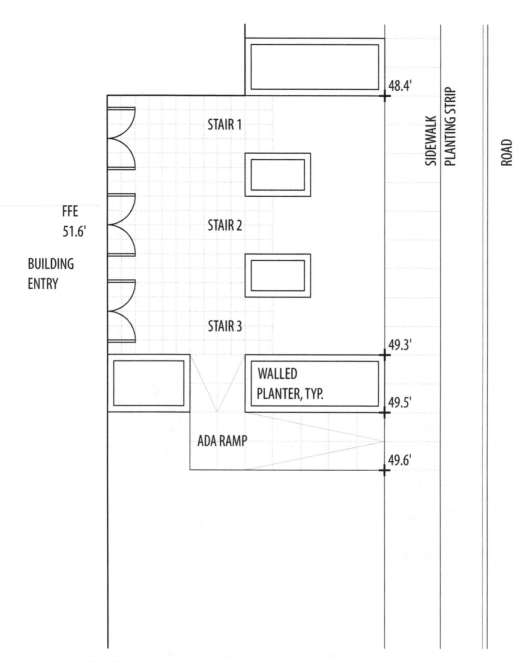

Figure 6.1. Plan for plaza grading

- Stairs shall have 12-in. treads and 6-in. risers and an extra tread width of landing at the top and bottom of each run to accommodate handrails.
- Draw in any walls that will be needed to construct your solution.
- Maximum slopes of walks shall be 5.0 percent.
- Provide your answer in spot elevations and rough grades.
- The scale of the plan is 1" = 10'-0".

6.3

Grade the basketball court located on the plan in Figure 6.3 using the following information:

- Assume that the curb, planter strip, and sidewalk are also new features and that the new bottom of curb elevation is represented by where existing grade hits the curb from the roadside.
- Ensure that water does not drain from the sidewalk over the basketball court.
- Provide swales as needed with a minimum slope of 2.0 percent.
- No slope shall be greater than 3:1.
- The court should be graded at 2.0 percent in both directions.
- The sidewalk and planter strip should be cross-sloped at 2.0 percent.
- The scale of the plan is 1" = 20"-0".

6.4

Grade the parking lot located on the plan in Figure 6.4 using the following information:

- Assume that all parking lot runoff is to be conveyed by swale, except where the water needs to pass under a roadway.
- The parking lot will have curb stops instead of full curbs to allow drainage to the swale system.
- Swales shall have a minimum slope of 2 percent.
- No slope shall be less than 2 percent.
- No slope shall be greater than 3:1.
- The design must meet the adjacent road alignment at the spot elevations provided.
- The scale of the plan is 1" = 40'-0".

6.5

On the plan in Figure 6.5, locate one 180' × 300' soccer field and three 120' × 60' tennis courts. Grade the fields using the following criteria:

- The tennis courts should have 15 ft clear between the courts and at least 12 ft clear around the courts.
- The soccer field should have 20 ft clear around its perimeter.
- The soccer field will be graded at 1.75 percent.
- The tennis courts will be graded at 1.0 percent.
- Provide swales as needed to ensure that no water flows across the fields from uphill.
- Swales will have a minimum slope of 1.0 percent.
- Tie any swales into the natural drainage of the site.
- No slope shall be greater than 3:1.
- The scale of the plan is 1" = 60'-0".

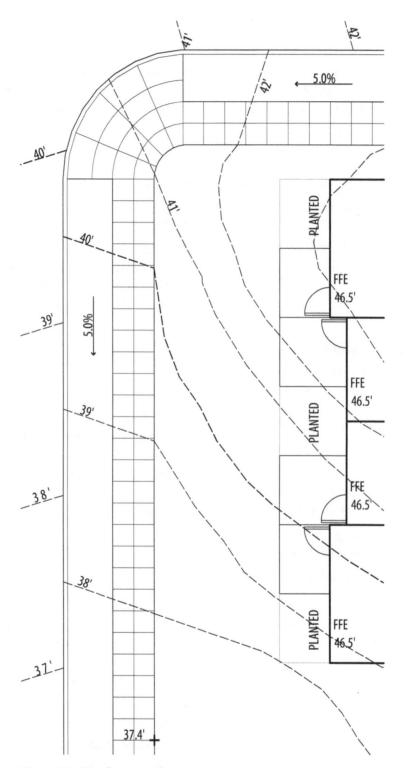

Figure 6.2. Plan for entry design

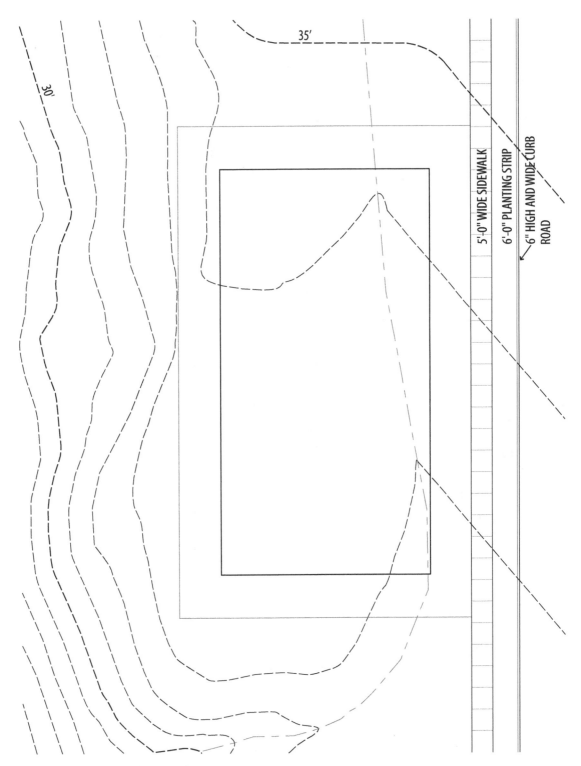

Figure 6.3. Basketball court plan

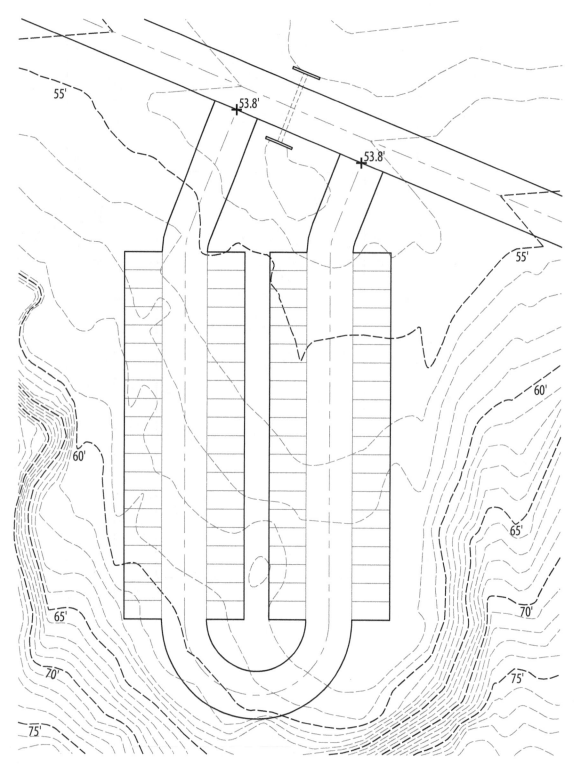

Figure 6.4. Parking lot plan

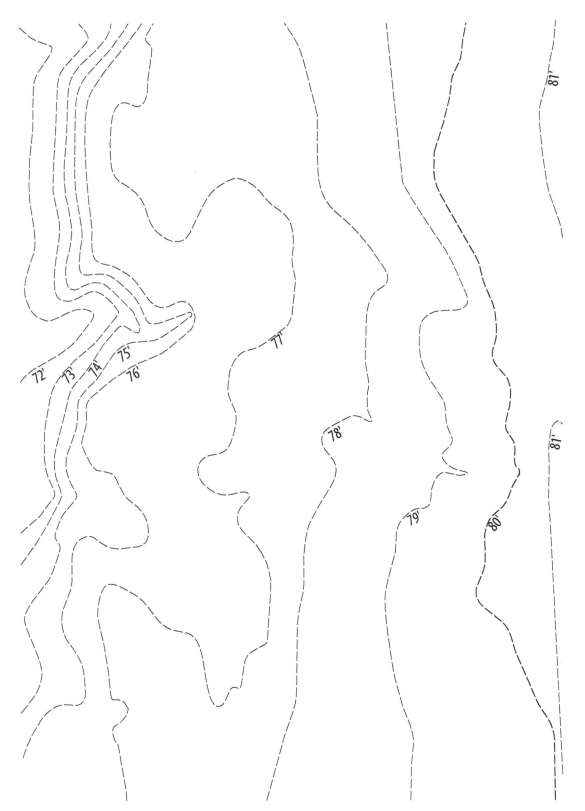

Figure 6.5. Sports courts plan

6.6

On the plan shown in Figure 6.6, use the spot elevation and slope information provided to grade the street to the interior of the spot elevations, assuming that grade slopes evenly between the spot elevations on the centerline of the road.

Once the road has been graded, design a plaza surrounding the building centered in the block using the following criteria:

- Slopes in the plaza are not to be less than 0.7 percent and not to exceed 2.5 percent.

- Provide positive drainage away from the building.

- The sets of doors opposite each other on the narrow dimension of the building must share the same finish floor elevation.

- In the long dimension of the building, grades should terminate evenly along the building, between the two sets of doors.

- Outside each set of doors, there should be a 5-ft landing that does not slope more than 2.0 percent.

- Use stairs, walls, and planters as necessary to achieve the above goals and tie into the elevations at the curb, understanding that the architect has requested that the plaza feel as open as possible.

- No slope shall be greater than 4:1 within any added planters.

- Draw 0.5-ft contours to get a clearer understanding of the topography.

- The scale of the plan is 1" = 40'-0".

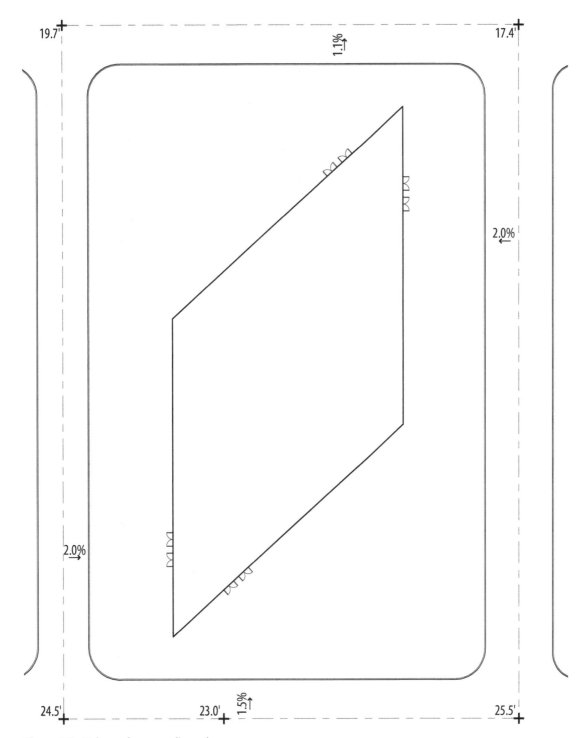

Figure 6.6. Urban plaza grading plan

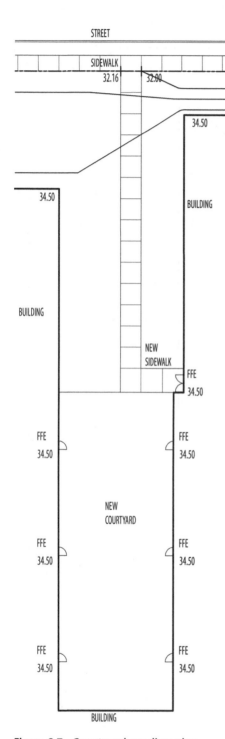

Figure 6.7. Courtyard grading plan

6.7

On the plan shown in Figure 6.7, use the following criteria to grade within the courtyard and out to the sidewalk, tying into existing contours:

- Slopes in the plaza are not to be less than 0.7 percent and not to exceed 2.0 percent.
- Provide positive drainage away from the building.
- Provide three (3) drain inlets within the courtyard and ensure that if one gets clogged, water will not back up into the building.
- In the long dimension of the building, grades should terminate evenly along the building, between the two sets of doors.
- The slope of the walk is to be between 1.5 and 5.0 percent with a cross slope between 1.5 and 2.0 percent.
- Use stairs, walls, and planters as necessary to achieve the above goals and tie into the elevations at the sidewalk.
- No slope shall be greater than 4:1.
- The scale of the plan is 1" = 30'-0".

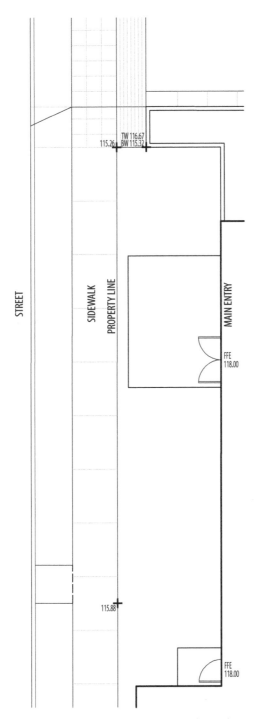

Figure 6.8. Urban streetscape grading plan

6.8

On the plan shown in Figure 6.8, provide stairs connecting the two entries to the sidewalk. Also provide a ramp from the main entry to the sidewalk. Assume that grade slopes evenly between the spot elevations along the sidewalk property line edge.

Grade the ramps, stairs and planting areas between the property line and the building using the following criteria:

- Slopes in the landings outside the entries are not to be less than 0.7 percent and not to exceed 2.0 percent.

- Provide positive drainage away from the building.

- Stairs risers are to be a minimum of 4" and a maximum of 7" tall.

- Stair treads are to be 14" wide.

- Ramp is to be a maximum of 8.3 percent.

- Show handrails at ramps with a slope greater than 5 percent slope.

- Ensure there is a 5' landing at the top and bottom of all stairs and ramps. The landing is to have a slope between 1.0 and 2.0 percent.

- Outside each set of doors, there should be a 5-ft landing that does not slope more than 2.0 percent.

- Provide room for 12" handrail overruns at top and bottom of stairs and ramps. Ensure these overruns do not protrude into plaza or landing spaces.

- Use stairs, walls, and planters as necessary to achieve the above goals and tie into the elevations at the sidewalk.

- No slope shall be greater than 4:1 within any planters.

- Show locations of barrier rails, where the change in grade is greater than 30".

- The scale of the plan is 1" = 10'-0".

CHAPTER 7

Questions

7.1

Visit the construction site you identified in Chapter 1. Observe how wind and water are impacting the soil on the site. Look at the foundations and footings that are being used and their relation to the landscape around them. How far away from these structures is the landscape impacted and the soil disturbed? Look at excavation for utilities. How steeply sloping are the sides of the utility trenches? If you happen to observe the site while new soil is being emplaced, what kind of machinery is being used for distribution and compaction of soil?

7.2

Use the flow diagram provided in Figure 7.1 to examine the soil at your home or other locations, where you have permission to disturb soil. What soil texture do you end up with using this guide? Looking closely at the composition, does your observation match the results from using the flow diagram?

7.3

What three factors play a part in determining how steeply a slope can be graded without the construction of a retaining wall or other grade change device?

7.4

Loads applied to foundations and retaining walls in addition to the earth being retained are called _____.

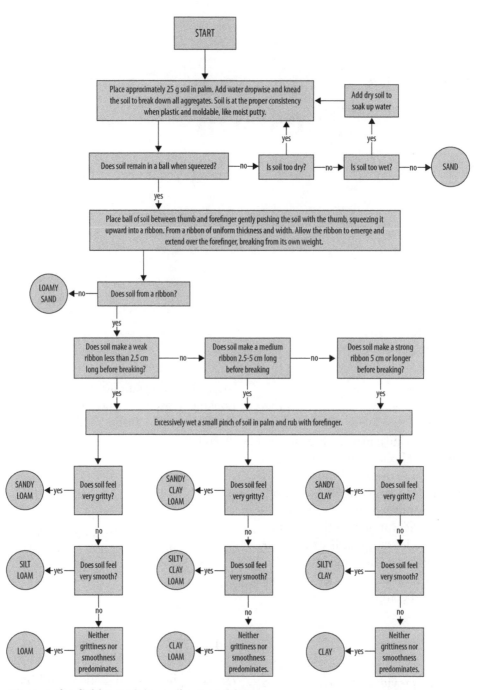

Figure 7.1. Diagram for field-examining soil composition

Source: This image is available from the USDA/NRCS website (http://soils.usda.gov/education/resources/k_12/lessons/texture/) and is modified from S. J. Thien, "A Flow Diagram for Teaching Texture by Feel Analysis," *Journal of Agronomic Education* 8 (1979): 54–55.

7.5

Give three examples of the types of loads described in question 7.4.

7.6

What are the four states of soil consistency, and what are the names of the limits that define the boundary between each state?

7.7

For the points located on the USDA soil textural triangle in Figure 7.2, what are the percentages of sand, silt, and clay for each (A-E)?

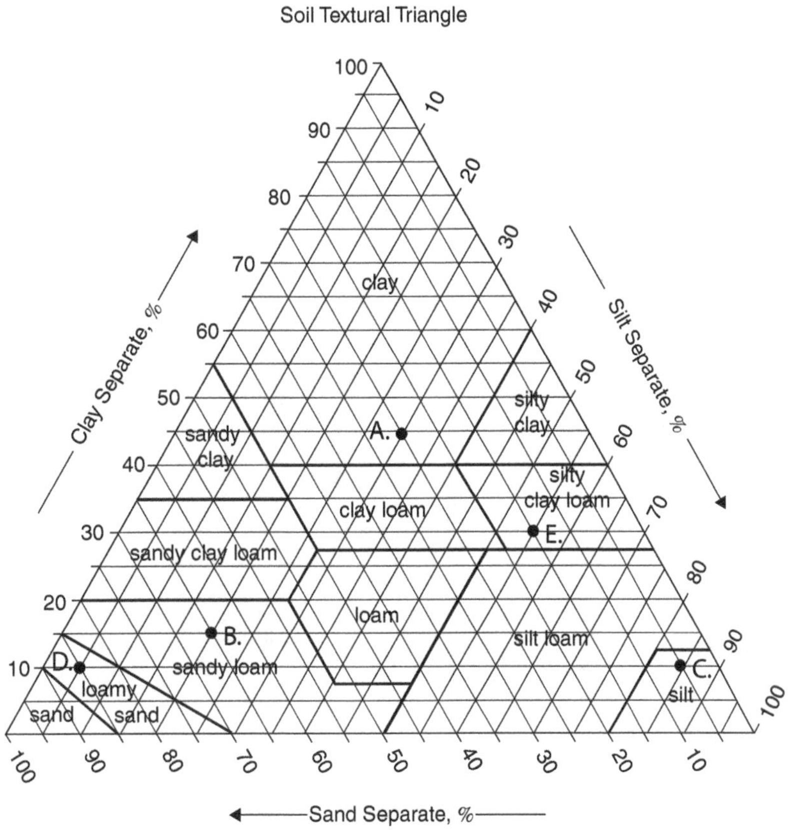

Figure 7.2. USDA soil textural triangle

7.8

Locate the following soil compositions on the USDA soil textural triangle in Figure 7.3. After locating the compositions, note their classifications.

A. 0% sand, 10% silt, 90% clay

B. 50% sand, 10% silt, 40% clay

C. 20% sand, 60% silt, 20% clay

D. 90% sand, 5% silt, 5% clay

E. 25% sand, 40% silt, 35% clay

7.9

When examining a site and determining the potential for soil to support a structure, the soil's _____ must be identified.

7.10

Name the two factors that affect the shear strength of a soil.

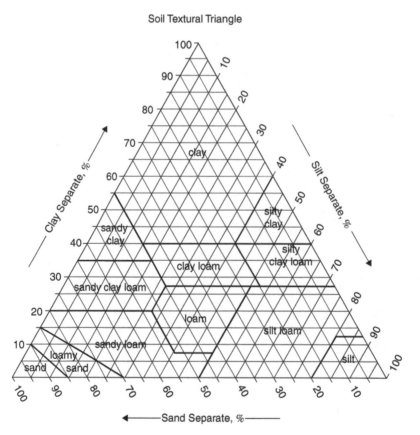

Figure 7.3. USDA soil textural triangle

7.11

What two types of soil particles are considered cohesionless?

7.12

Slope failure occurs because of either (increased/decreased) stress or (increased/decreased) strength brought about by natural or human-induced activity.

7.13

(True/False) An increase in moisture content in soil can result in both decreased strength and increased stress of the soil.

7.14

Soil with high proportions of _____ particles is prone to shrinking and swelling.

7.15

A soil that does not contain midrange-size stone is considered _____ and is a major component in the creation of _____ soil.

7.16

What natural environment is often considered a model for the design of green roof soil medium?

7.17

What are the four general functions for which geotextiles are employed?

7.18

Scarifying the landscape after soil has been compacted improves what two things that increase the soil's ability to support plant life?

7.19

What are the different compositions of the three (3) types of structural soil discussed in the chapter?

7.20

What two functions are structured soil volumes designed to perform?

7.21

What are the eight (8) steps in the grading sequence, in the order in which they are performed?

7.22

A _____ is a uniform layer of fill installed in a specified thickness.

CHAPTER **8**

Questions

8.1

Use the station information provided in Table 8.1 and use the average end area method to determine the amount of cut or fill required for this segment of road improvements. Indicate any amount of cut or fill required for export or import in yd^3.

8.2

Use the contour area method to determine the amount of cut or fill required for the grading in the figure shown in the answer to question 5.10.

Table 8.1. Station Area Data

Station	Area of Cut (ft^2)	Area of Fill (ft^2)
0+00	0.0	10.0
0+50	0.0	75.3
1+00	−5.5	124.3
1+50	−24.7	157.6
2+00	−55.8	235.4
2+50	−78.6	175.3
3+00	−125.5	111.4
3+50	−178.9	25.3
4+00	−155.5	13.1
4+50	−107.7	0.0
5+00	−93.5	0.0
5+50	−67.7	0.0
6+00	−38.2	5.7
6+50	0.0	0.0

Create a contour area measurement table to organize your measurements. The image in the figure is drawn at a scale of 1" = 20'-0". Your answer should be provided in cubic yards of cut and fill.

Tracing this plan in AutoCAD at full scale may make the measuring of contour areas more convenient. Alternatively, a planimeter can be used to measure contour area. Before measuring the contour areas, establish a line of no cut–no fill.

8.3

Use the contour area method to determine the amount of fill required for the grading in the figure shown in the answer to question 5.7. Create a contour area measurement table to organize your measurements. The image in the figure is drawn at a scale of 1" = 40'-0". Your answer should be provided in cubic yards of cut and fill.

Tracing this plan in AutoCAD at full scale may make the measuring of contour areas more

convenient. Alternatively, a planimeter can be used to measure contour area. Before measuring the contour areas, establish a line of no cut–no fill. Assume that the no cut–no fill line falls 5 ft beyond the proposed contours, when the next adjacent contour is undisturbed.

8.4

Use the borrow pit method to determine the amount of cut required to excavate for the basement of the building proposed in Figure 8.1 to a depth of 26.9 ft. The image is drawn at a scale of 1" = 20'-0". Indicate the amount of cut required in cubic yards.

8.5

Use the borrow pit method to determine the amount of cut or fill required for grading the terrace

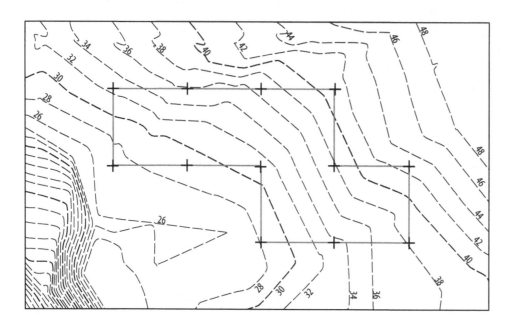

Figure 8.1. Grading plan for basement

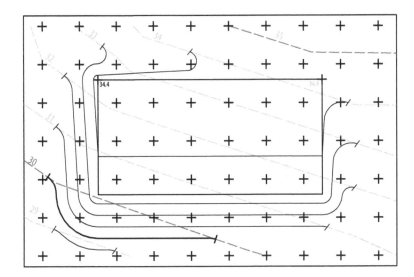

Figure 8.2. Grading plan for terrace

in Figure 8.2 (which you may recognize as the figure in the answer to question 5.9). The image is drawn at a scale of 1" = 20'-0". Use the 10' × 10' grid provided for your borrow pit calculations.

Adjust the volumes using the following design criteria:

a. Depth of existing topsoil is 6 in.

b. Depth of proposed topsoil replaced only in disturbed areas is 8 in.

c. Depth of proposed pavement at the terrace is 4 in.

Indicate any amount of topsoil and cut or fill required for export or import. Provide all answers in cubic yards.

CHAPTER 9

Questions

9.1

Look up regional resources that describe the nature of storm water's interaction with your regional climate. The EPA has divided the United States into different regions and, for each region, typically provides some discussion of storm water and efforts to develop ways of managing it that are appropriate to the particular region (www.epa.gov/epahome/regions.htm).

The Center for Watershed Protection is also a clearinghouse for the most up-to-date storm water management information, much of which is offered free on its website (www.cwp.org).

9.2

The increased amount of impervious surface that results from development changes the surface of the landscape. What are the important changes that occur to the landscape surface during development?

9.3

List four environmental impacts that result from the changes in surface characteristics mentioned in question 9.2.

9.4

(True/False) Higher velocities coupled with increased imperviousness may also result in reduced stream flow during extended dry periods.

9.5

List three ways that stream geometry changes as a result of the hydrologic changes caused by development, particularly the increased volume and velocity of surface runoff.

9.6

Water quality is also impacted negatively as a result of development. Describe two of the ways that developed landscapes affect water quality.

9.7

An integrated approach to storm water management addresses what three factors altered by landscape development?

9.8

Degradation of water quality in streams is known to take place when the impervious surface coverage of a drainage area approaches approximately _____ percent.

9.9

What are two ways to manage the impact of developing parking lots through design?

9.10

What are the seven functions being incorporated into the design of modern storm water management systems to mitigate storm water quantity and quality impacts brought by impervious surfaces in the built environment?

9.11

Name four types of treatment being designed into modern storm water management systems.

9.12

_____ is the storage of rainwater on the surfaces of a planting.

9.13

Soluble nutrients, metals, and organics are removed from storm water runoff through a process called _____.

CHAPTER 10

Questions

10.1

Look up your local municipal code as it pertains to storm water management. Is there any discussion of the storm water management structures discussed in the chapter? Is there any encouragement to use the BMPs outlined in the textbook? Are there particular places where they are deemed appropriate and other places where they are not desirable?

10.2

Visit a green roof or planting on structure. Look at how the plants and soil volume are contained. How close to building walls do the plantings get? Are there particular plants that appear to be thriving better than others? Look at the texture and makeup of the soil. Is anything being used to keep the soil in place, such as a fiber mat or stone mulch?

Photograph the green roof and upload the photos with your observations to the Site Engineering for Landscape Architects Facebook page.

10.3

Visit local storm water management BMPs.

- Can you make any observations about how successfully they are functioning?
- Can you identify the different elements that make up each BMP as described in the textbook?
- Are they designed to invite human interaction, or are they cordoned off from the public?

Photograph the BMPs and upload the photos with your observations to the Site Engineering for Landscape Architects Facebook page.

10.4

What are the two types of culvert, and what types of cross sections can they be purchased in?

10.5

Generally, a single catch basin may be used to cover what area of paved surface drainage?

10.6

What is the primary difference between a catch basin and a drain inlet?

10.7

Best management practices for storm water management can generally be described as providing what three functions?

10.8

What particular site conditions are important to analyze when selecting a BMP?

10.9

Why is maximization of the length of the path of flow between inlet and outlet an important criterion in the design of retention ponds?

10.10

Name three techniques for surface detention of storm water on a site.

10.11

Name three techniques for subsurface detention of storm water on a site.

10.12

What practical limitations of a landscape does a designer need to be concerned with when considering the use of an infiltration structure?

10.13

Name four functions that a constructed treatment wetland might perform.

10.14

What three elements must be included in any rainwater harvesting system?

10.15

What differentiates extensive and intensive green roof systems?

10.16

What vital function does the drainage and retention layer provide in a green roof system?

10.17

Green roof systems that have been designed to provide habitat are called _____.

10.18

What is the conceptual relationship between bio-retainment and bioretention?

10.19

What are the two goals of Net Zero Water as defined by the Living Building Challenge?

11
Questions

11.1

Look up your local municipal code as it pertains to erosion and sediment control in construction. Many municipalities now require a temporary erosion and sediment control plan for construction of a certain scale. Is a threshold listed, or are best practices recommended?

11.2

Visit the construction site you identified in Chapter 1. Look for erosion control structures described in the textbook that have been erected on-site. Do you see any potential conflicts between the existing landscape and ongoing construction processes? For example, is soil being eroded away from or deposited onto existing vegetation?

11.3

The cycle of erosion and sedimentation involves what three steps?

11.4

What are the four natural processes that can be responsible for causing soil to move, starting the process of erosion?

11.5

What are the four primary factors that determine a soil's potential for erosion?

11.6

Organize the following types of soil particles in order from the greatest erosion potential to the least erosion potential: well-sorted gravels and gravel-sand mixes, clay, and silt and fine sand.

11.7

With respect to vegetative cover, the goal of most site development projects should be _____.

11.8

What are the five basic principles for minimizing disturbance while developing a site?

11.9

What are the five types of runoff control measures?

11.10

Provide a definition of soil bioengineering.

CHAPTER 12

Questions

12.1

What assumptions does the Rational method make that limit its application to large landscape areas?

12.2

When values of the runoff coefficient, C, get closer to 1, they are becoming (more/less) pervious.

12.3

Rainfall intensity, i, is a measurement that corresponds to which two factors of a rainfall event?

12.4

(True/False) A 20-year design storm will happen once every 20 years in a particular location.

12.5

Define time of concentration (T_C).

12.6

Using the nomograph for overland flow in Figure 12.1, determine the inlet concentration time in minutes.

A. 600-ft length of drainage on bare soil at a 2 percent slope

B. 300-ft length of drainage on woodland at a 5 percent slope

C. 400-ft length of drainage on a paved surface at a 1 percent slope

12.7

Determine the peak runoff rate for a drainage area consisting of the following:

- 1.2 acre of gravel roadway with a runoff coefficient of 0.70

- 1 acre of rooftop with a runoff coefficient of 0.95

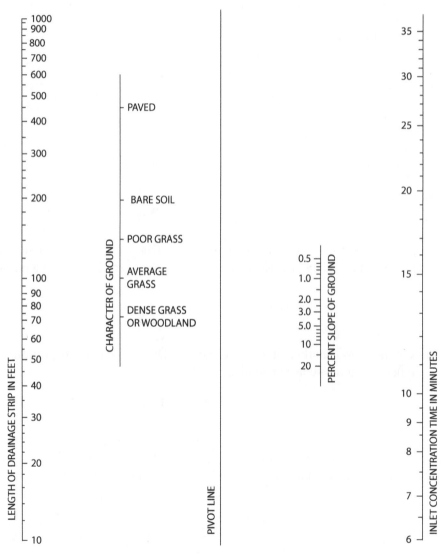

Figure 12.1. Nomograph for overland flow time

- 2.4 acres of woodland with a runoff coefficient of 0.50
- 4.2 acres of lawn with a runoff coefficient of 0.10

12.8

For a new project site, the topographic and land use maps have been examined, and it has been determined that the path from the hydraulically most remote point of the drainage area to the point of concentration takes the following course:

- 800 ft of paved surface at a 1.5 percent slope
- 140 ft of poor grass surface at a 0.9 percent slope
- 60 ft of dense grass at a 6.0 percent slope
- 1,600 ft of stream flow at a 1.0 percent slope. Use the nomograph in Figure 12.2 to obtain the value for this element.
- Find the time of concentration for this area.

12.9

What type of elements can be sized using calculations of peak runoff rates?

12.10

What information must be determined to size storage ponds and reservoirs?

12.11

The drainage area of the project site introduced in question 12.8 contains 14 acres of a cemetery with a runoff coefficient of 0.25, 6 acres of railroad yards with a runoff coefficient of 0.35, 20 acres

of light industry with a runoff coefficient of 0.80, and 26 acres of unimproved urban landscape with a runoff coefficient of 0.30. Determine the peak storm runoff rate for a 50-year frequency if the site is located in Trenton, New Jersey. Use Figure 12.3 as needed for your calculations.

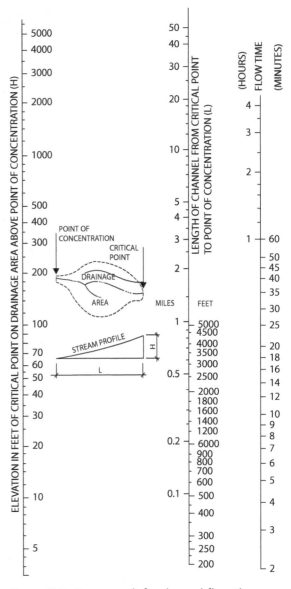

Figure 12.2. Nomograph for channel flow time

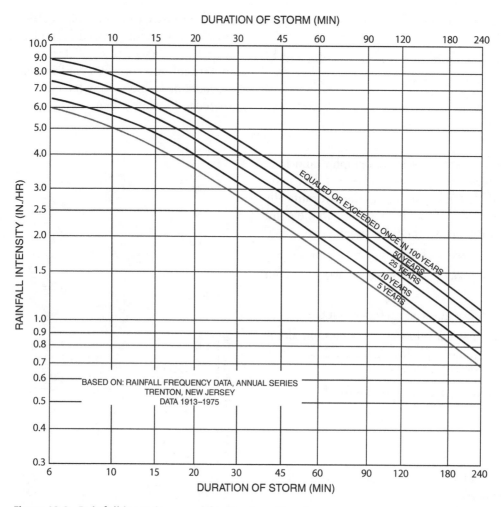

Figure 12.3. Rainfall intensity curves for Trenton, New Jersey.

12.12

For a type A hydrograph, the duration of the storm event is _____ the time of concentration.

For a type B hydrograph, the duration of the storm event is _____ the time of concentration.

For a type C hydrograph, the duration of the storm event is _____ the time of concentration.

12.13

Determine the 100-year peak flow rate for a 19-ac drainage area using the MRM, given the following characteristics:

An overland flow path of 600 ft of a poor grass surface at a 2.0 percent slope. The project is located in Trenton, New Jersey, and is on clay and silt loam soils in woodland.

Use the figures already provided in the chapter as needed, as well as Table 12.1.

Table 12.1. Recommended Antecedent Precipitation Factors (C_A)

Frequency	C_A
2 to 10	1.0
25	1.1
50	
100	−1.2
1.25	

Source: American Public Works Association (1981). "Practicing in Detention of Urban Storm Water." Special Report 43.

12.14

Develop sketch hydrographs for the area described in question 12.13 for the following durations: 10, 20, and 40 minutes.

12.15

The area from question 12.13 is to be developed as apartments, and its time of concentration will be reduced to 10 minutes. Determine the 100-year peak runoff rate by the MRM and using the figures and tables already provided in the chapter. In cases where a range of values is listed in tables, use the higher value in the range.

12.16

What will be the required maximum (critical) storage volume for the area in question 12.15 if the outflow rate of a 100-year event cannot exceed the predevelopment rate? What will be the duration of the "critical" storm?

CHAPTER 13

Questions

13.1

In using the NRCS method for estimating runoff, inches of precipitation are translated into inches of runoff using a _____, which is based on soils, land use, impervious areas, interception by vegetation and structures, and temporary surface storage.

13.2

The TR-55 manual separates flow into three different and distinct processes. What are they?

13.3

Search for and download the Technical Release 55: Urban Hydrology for Small Watersheds (Issued June 1986) from the NRCS website.

In addition to the above resource, locate soil information for your local county, usually available online from the local county-level municipality.

Using these materials, look up a few of your local soils in Appendix A ("Hydrologic Soil Groups") in the TR-55 manual. Do the soils in your area appear to fall into typical groups?

13.4

In the TR-55 manual, look in Appendix B ("Synthetic Rainfall Distributions and Rainfall Data Sources") and identify the rainfall distribution category for your region. Read about how these different types correspond with different climatic conditions. Look up the rainfall expected with different design storms in your local region. How does the rainfall differ between your local region and other places you have visited or lived?

13.5

Look up Appendix D ("Worksheets") in the TR-55 manual. The worksheets in this appendix correspond to the different calculations covered in the textbook. Use this new method for organizing information and try to replicate the examples provided in the textbook.

Also review the various chapters in the manual that lay out the process of making the calculations when you are trying to replicate the example problems. Pay particular attention to the limitations listed for each set of calculations.

13.6

Search for and download the WinTR-55 computer program and the WinTR-55 manual from the NRCS website and install it on your home computer. It is a free program.

Once you have installed the program, follow the step-by-step example in Appendix A in the WinTR-55 manual. Once you have done this, attempt to input the examples from both the textbook and the TR-55 manual. Does the program come up with different answers than the original sources?

CHAPTER 14

Questions

14.1

Name four factors that impact the selection of an appropriate drainage system.

14.2

What are the three primary functions of a storm drainage system?

14.3

(True/False) It is illegal to increase or concentrate flow across landscape outside of your project's property line.

14.4

Swales and pipes generally (increase/decrease) in size as they progress toward the outlet point.

14.5

In a swale, what are three different ways to reduce the velocity of water through design?

14.6

Reducing the velocity of water flowing in a swale results in _____.

14.7

What three factors determine the permissible maximum design velocity of a swale?

14.8

What is the flow velocity of a parabolic swale if the peak rate of runoff is 49 ft^3/s and the cross-sectional area is 14 ft^2?

14.9

Design a parabolic swale to carry 91 ft^3/s of runoff at a slope of 3.5 percent. Its permissible velocity is 4 ft/s, and its roughness coefficient is 0.05.

14.10

A flow of 4.5 ft^3/s must be conducted by a concrete circular pipe ($n = 0.015$). The gradient must be 1.2 percent due to site conditions. Determine the required diameter of the pipe and the flow velocity in the pipe.

14.11

Use the information provided in Table 14.1 to determine the pipe sizes required for the subsurface drainage system illustrated in Figure 14.1.

14.12

Given the following assumptions, what is the minimum size roof needed to capture enough water to fill the cistern in one month?
    Cistern size: 4,500 cubic feet
    Average monthly rainfall: 4.3 inches
    Efficiency of capture: 80 percent

14.13

Given the following assumptions, what size cistern is needed to get the planting through a four-week drought?
    Planted area: 500 square feet
    Weekly irrigation required: 0.75 inches of water per square foot of planted area

Table 14.1. Maximum acreage[a] drained by various pipe sizes: clay or concrete pipe ($n = 0.011$, DC = $^3/_8$ in. /24 hr.)

Pipe size (in.)	Slope (%)									
	0.1	0.2	0.3	0.4	0.5	0.6	0.7	0.8	0.9	1.0
4	4.51	6.38	7.82	9.03	10.1	11.1	11.9	12.8	13.5	14.3
5	8.19	11.6	14.2	16.4	18.3	20	21.7	23.2	24.6	25.9
6	13.3	18.8	23.1	26.6	29.8	32.6	35.2	37.6	39.9	42.1
8	28.7	40.5	49.6	57.3	64.1	70.2	75.8	81.1	86	90.6
10	52	73.5	90	104	116	127	138	147	156	164
12	84.5	120	146	169	189	207	224	239	254	267

[a]Reduce these acreages by one-half for a ¾-in. DC.

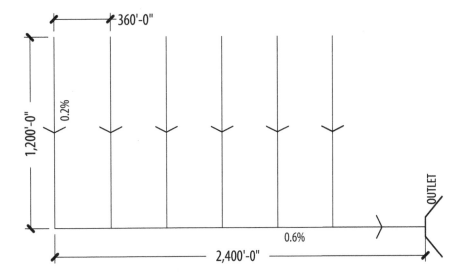

Figure 14.1. Piping plan

CHAPTER 15

Questions

15.1

Visit the construction site you have chosen to observe, looking for evidence of layout of elements to be constructed. Are there stakes marking out grades or the alignment of elements to be constructed, or outlines of elements painted on the ground? Are there certain features that require one or the other method of location? How are measurements being taken in the field? Is a surveyor working on the construction?

15.2

The purpose of a layout plan is to establish _____ position, orientation, and extent of all proposed construction elements. By contrast, _____ position is established by the grading plan.

15.3

The majority of site dimensions are _____ dimensions.

Provide an example of this type of dimension.

15.4

Name an item not typically dimensioned on a layout plan.

15.5

The measurement 34.6 ft implies a preciseness of ±_____.

15.6

What is the starting point of a layout plan called?

15.7

What is wrong with the following expression of a layout dimension?
    0'-11"

15.8

_____ is the most appropriate method of layout when site elements are located orthogonal to property lines or proposed buildings.

15.9

What system of dimensioning is best used on curvilinear elements that may not require a high degree of accuracy?

15.10

The plan in Figure 15.1 is drawn at a scale of 1" = 10'-0". Use baseline dimensioning to locate the planting bed in relation to the surrounding sidewalk. The POB has been located on the plan. Feel free to use other site features to guide the placement of your dimensions.

15.11

The plan in Figure 15.2 is drawn at a scale of 1" = 20'-0". Use perpendicular offsets to locate the path, the center of trees, and their surrounding planting beds. The POB has been located on the plan.

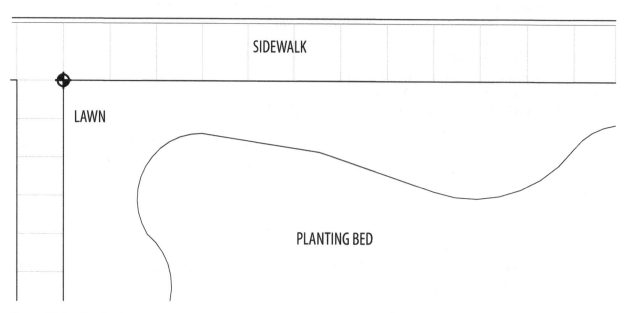

Figure 15.1. Plan for layout

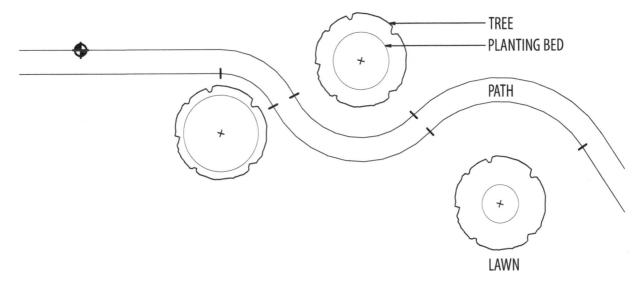

Figure 15.2. Plan for layout

15.12

The plan in Figure 15.3 is drawn at a scale of 1" = 10'-0". Use perpendicular offsets to locate the path, the center of trees, and their surrounding planting beds. Select a POB.

Where appropriate, reinforce the design intent by noting sets of joints that should be equally spaced, by noting the number of even units followed by "EQ." For example, if five joints divide a concrete panel that is 6'-0" wide into six parts, a dimension noting "6 EQ" should be nested underneath the dimension noting 6'-0".

15.13

What are the two advantages to using the coordinate system for laying out information?

15.14

A bearing is an _____ angle, measured off of the _____ axis or meridian.

15.15

What three pieces of information are needed to define an arc?

15.16

What factors affect the degree of accuracy possible with a GPS receiver?

15.17

How many satellites must be perceptible by a GPS receiver in order to be able to obtain a location in three dimensions?

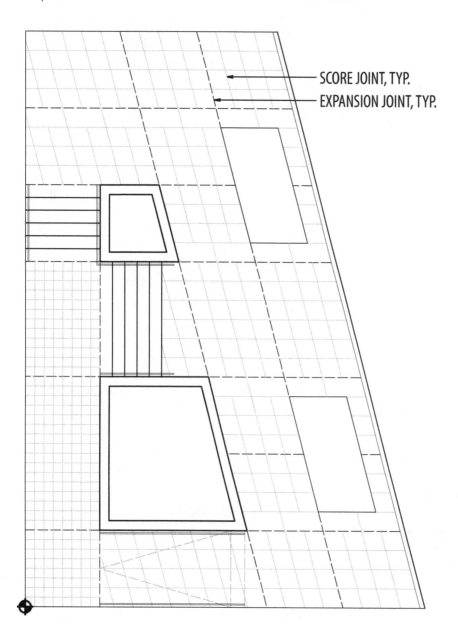

SCORE JOINT, TYP.
EXPANSION JOINT, TYP.

Figure 15.3. Plan for layout

CHAPTER 16

Questions

16.1

Visit the roads you have chosen to observe. What role does the horizontal alignment play in your experience of the roads? Is your visibility obstructed at turns in the road? Can you perceive a superelevation on the highway segment you chose?

16.2

What are the two basic components of a horizontal road alignment?

16.3

Name and sketch the three types of curves commonly used in horizontal road alignment.

16.4

Locate the following abbreviations for elements of a circular curve on the plan shown in Figure 16.1:
T, I, R, L, C, O, PI, PT, PC

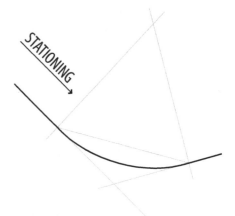

Figure 16.1. Curve plan

16.5

What do the different abbreviations mentioned in question 16.4 stand for?

16.6

(True/False) For simple horizontal curves, the distance from PC to PI and from PI to PT is always equal.

16.7

(True/False) A new road stationing system is usually started for each new road at the curb line of an existing road.

16.8

The rise of the outer edge of pavement relative to the inner edge at a curve in the highway, expressed in feet per foot, intended to overcome the tendency of speeding vehicles to overturn when rounding a curve, is called _____. Its value should not exceed _____ under most conditions, or _____ if snow and ice are a local problem.

16.9

Draw the figure defined by the following bearings at a scale of 1" = 60'-0":

- Bearing A to B travels 130' S32°14'W.
- Bearing B to C travels 240' N78°56'E.
- Bearing C to D travels 178.1' N68°58'W.

16.10

Given the following information, draw the bearings and curves and station the length of the horizontal alignment centerline at 100-ft intervals. Determine the radii, tangents, lengths, and chords for each curve.

- Bearing A to B travels 720' S80°7'W.
- Bearing B to C travels 640' S18°44'E.
- Bearing C to D travels 700' S63°49'W.
- Bearing D to E travels 600' S40°50'E.
- Curve 1 has a radius of 195'.
- Curve 2 has a tangent of 276.49'.
- Curve 3 has a radius of 235'.

16.11

Design a road to connect road 1 and road 2 in Figure 16.2 using at least one curve with a minimum radius of 450'-0" and a 40-mph design speed. The new road should intersect each existing road at a 90° angle. Determine the radii, tangents, lengths, and chords for each curve used. Draw in the centerline of the new road, locating 100-ft stations along its length. The scale of the figure is 1" = 100'-0".

16.12

Determine the radius, tangent length, length of curve, and chord length for a 20° curve with an included angle of 45°00'.

16.13

Compute the superelevation and runoff distances for a 375-ft radius curve for a road with a 40-mph design speed in a climate in which snow and ice are not a problem. The road crown is 0.25 in./ft.

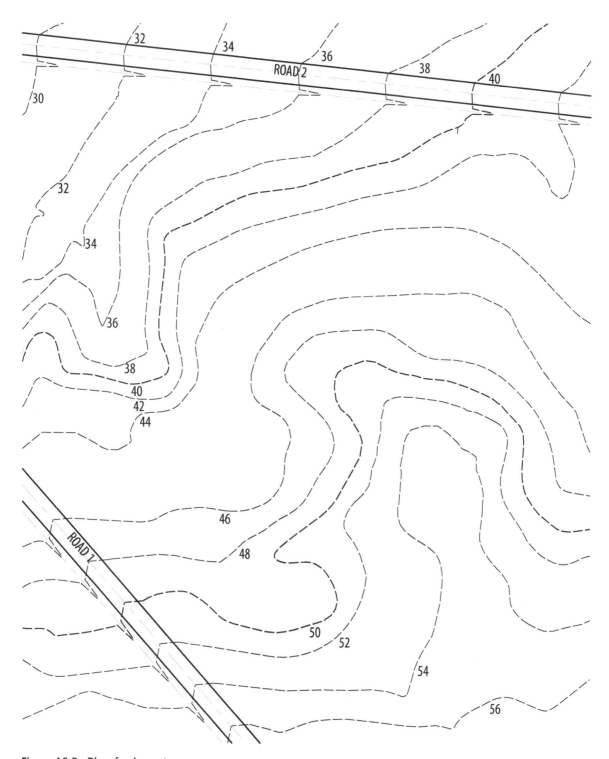

Figure 16.2. Plan for layout

CHAPTER 17

Questions

17.1

Visit the roads you have chosen to observe. What role does the vertical alignment play in your experience of the roads? Is your visibility obstructed by the vertical profile of the road at any point? What relationships to adjacent uses does the vertical alignment create?

17.2

Tangent lines for horizontal curves are _____ lines in the horizontal plane, whereas tangent lines for vertical curves are _____ lines in the vertical plane.

17.3

Name and sketch the four types of curves commonly used in vertical road alignment.

17.4

Locate the following abbreviations for elements of a vertical curve on the curve section in Figure 17.1:
$e, l, L, x, y,$ PVI, BVC, EVC

17.5

What do the abbreviations mentioned in question 17.4 stand for?

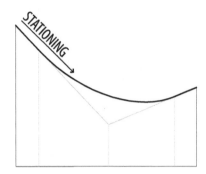

Figure 17.1. Curve section

17.6

(True/False) The formulas and calculations used to determine road curves can also be used to lay out any other curved features, including walls, fences, and pathways.

17.7

A vertical curve with a −2.1 percent slope coming into the PVI and a −5.7 percent slope leaving the PVI has an algebraic difference of _____ percent.

17.8

From the information given, calculate curve elevations for an equal tangent curve at each 50-ft station (0 + 00, 0 + 50, etc.) and determine the station points for BVC, EVC, and the low point. Present your information in a table, including columns of information for the station, the point's name if it has one, the tangent elevation, the tangent offset, and the curve elevation.

- The slope of the entering tangent is −1.7 percent.

- The slope of the exiting tangent is +6.1 percent.

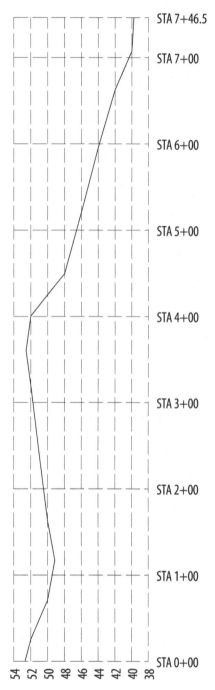

Figure 17.2. Section for creation of vertical curve profile

- The total length of the curve is 350 ft.
- The elevation at the PVI is 49.2 ft.
- The BVC is located at station 0 + 50.

17.9

Create equal tangent vertical curves using the following information and the section shown in Figure 17.2. Locate BVCs, EVCs, HPs, LPs with stations, and elevations. The horizontal scale is 1" = 100'-0". The vertical scale is 1" = 10'-0".

- The first tangent point is located at station 0 + 00, elevation 52.6'.
- The PVI of the first curve is located at station 1 + 117, elevation 49.1'.
- The PVI of the second curve is located at station 3 + 60.3, elevation 52.6'.

- The last tangent point is located at station 7 + 46.5, elevation 39.7'.
- The first vertical curve will have a length of 150'.
- The second vertical curve will have a length of 200'.

17.10

The vertical curves calculated in question 17.9 are one possible solution for vertical curves associated with the horizontal alignment created in question 16.11. Using the section created in question 17.9, grade the alignment shown in Figure 17.3 for a 12-ft-wide road, using swales and culverts as necessary. After implementing the new grading, examine how the vertical curves might be changed to better fit the alignment to the topography. The scale of the figure is 1" = 100'-0".

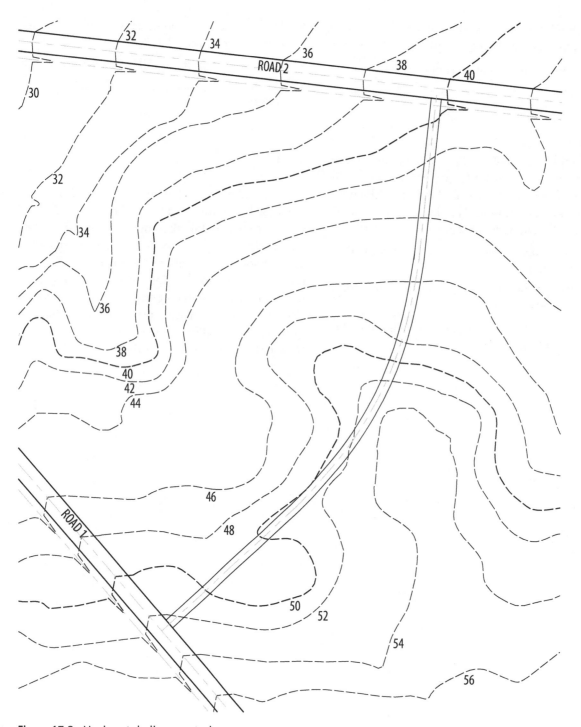

Figure 17.3. Horizontal alignment plan

ANSWERS

For each chapter of questions, there is a corresponding chapter of answers. Answers for all questions, except those designed as observations, are provided in this section. Question numbers for the observation questions are entirely omitted from the Answers section. In the case of the more complex graphic questions, the answer provided is typically one of many possible solutions. Valid alternative solutions can be created as long as all the criteria mentioned in the question have been satisfied.

CHAPTER 2

Answers

2.7

Increase in impervious surface in construction results in (**increased**/~~decreased~~) fluctuation of water levels in streams, ponds, and wetlands and (**increased**/~~decreased~~) potential for flooding.

2.8

Grade change should be kept outside of the drip line of vegetation to best protect that vegetation. Some vegetation is especially sensitive to disturbance, particularly compaction, even outside of its drip line. It is best to consult with a licensed arborist to help determine the best course of action when building in close proximity to mature vegetation.

2.9

Any of the following answers are appropriate: loose silts; soft clays; fine, water-bearing sands; and soils with high organic content, such as peat.

2.10

The length and degree of slopes. Depending on the existing soil structure on the site, the potential for erosion increases with either or both of these factors.

2.11

Any of the following answers are appropriate: maximum allowable cross slope for public sidewalks, maximum height of street curbs, acceptable riser/tread ratios for stairs, maximum number of stair risers without an intermediate landing, maximum slope for handicapped ramps, slope protection, drainage channel stabilization, retention structures, and vegetative cover.

2.12

The best starting point is the elevation at the point of connection to the existing drainage system. This should allow in most cases for the design of a system that drains by gravity, which avoids the expense and future maintenance of a pump to help the system function.

2.13

Pole, or pier, construction can provide a great amount of flexibility, but that flexibility may also come at a greater expense. Design intent and budget are a constant juggling act in getting your ideas constructed.

2.14

1. Terraces and sitting areas: **b.** 1 to 2%. This permits chairs and benches to be arranged so that occupants do not feel they are on a slope but still provides enough slope to allow for drainage.

2. Planted banks: **f.** up to 50%. The roots of the vegetation add structure to the slope that limits erosion, and the leaf cover helps slow down the rain on its way to the ground, so steeper slopes can be maintained.

3. Playfields: **c.** 2 to 3%. With large footprints requiring relatively low slopes, playfields may often work toward the upper limit to fit into their surroundings while having a limited impact on play.

4. Public streets: **e.** 1 to 8%. Because of the wide range of conditions that public streets must traverse, this range may appear a bit wider than you might expect.

5. Parking areas: **d.** 1 to 5%. Parking lots are meant to be surfaces traversed by people with all different capacities for mobility. It is difficult to ensure that slopes below 1% will drain without ponding, and many people find slopes greater than 5% difficult to traverse without a handrail.

6. Game courts: **a.** 0.5 to 1.5%. These paved surfaces often require finer tolerances so that game play is not affected while providing for drainage. Exterior tennis and basketball courts are often found with areas where water pools on their surface because these fine tolerances were poorly constructed to begin with or have had minor areas of settlement in the subgrade.

2.15

This is referred to as providing "positive drainage." This can generally be stated as a desired goal when setting elevations of building entrances, though there will be exceptions, especially in landscapes with steeply sloping topography.

2.16

The property line defines the boundary of where grades can be changed. In other words, all grading must meet existing grades within the boundary provided by the property line.

CHAPTER 3

Answers

3.6

The figure shows a 20-ft contour interval. The 100-ft contours are labeled, and there are four intermediate contours drawn between the labeled contours.

3.7

A (~~concave~~/**convex**/~~uniform~~) slope is one in which the slope gets progressively steeper moving from higher to lower elevations.

3.8

The figure shows a natural landscape, but the contours drawn to represent it divide. Contours never divide or split in the natural landscape.

3.9

Only two contour lines are required to indicate a three-dimensional form and direction of slope.

3.10

The steepest slope is (~~parallel~~/**perpendicular**) to a contour line.

3.11

Figure 3.1A shows the result of taking a section at the cut line indicated.

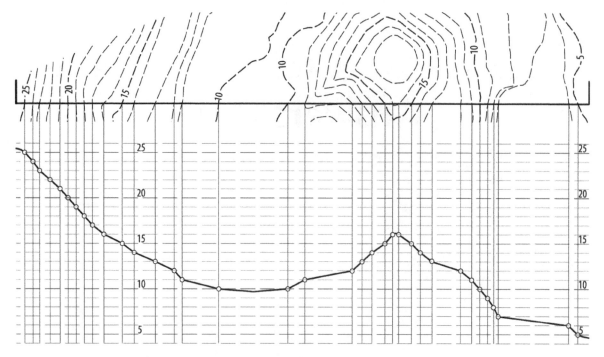

Figure 3.1A. Section resulting at the cut line indicated

3.12

Figure 3.2A shows the correct locations for the contour signatures.

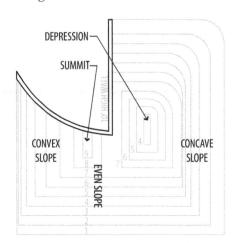

Figure 3.2A. Contour signatures for the contour plan

3.14

On the image shown in the figure:
A = convex slope.
B = concave slope.
C = even slope.
D = valley.
E = ridge.
F = summit.

CHAPTER 4

Answers

4.1

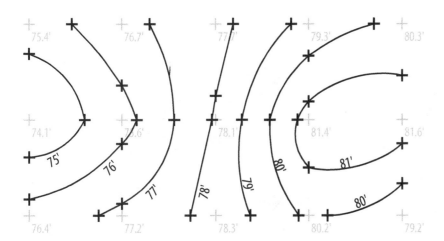

Figure 4.1A. Solution to interpolation

4.2

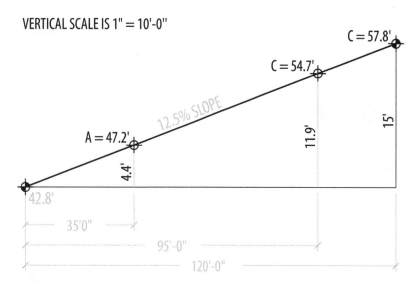

VERTICAL SCALE IS 1" = 10'-0"

C = 57.8'

C = 54.7'

A = 47.2'

12.5% SLOPE

4.4'

11.9'

15'

42.8'

35'0"

95'-0"

120'-0"

Figure 4.2A. Solution to slope triangle

4.3

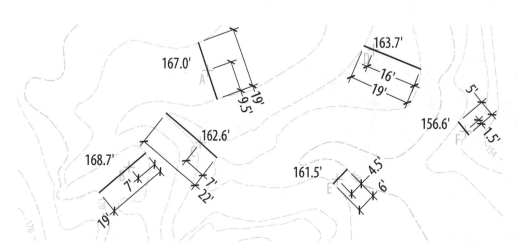

167.0'

163.7'

16'

19'

156.6'

5'

1.5'

168.7'

162.6'

7'

7'

22'

19'

161.5'

4.5'

6'

Figure 4.3A. Solution to interpolation

4.4

Slopes given the measurements shown in the figure below:

AC = 19.1%

AD = 6.1%

BC = 24.4%

CE = 2.4%

CF = 7.4%

DF = 18.2%

EF = 11.1%

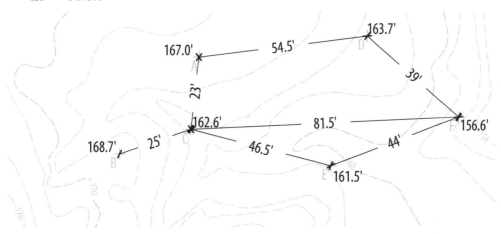

Figure 4.4A. Measurements to obtain slopes

4.5

Figure 4.5A. Solution to slope problems

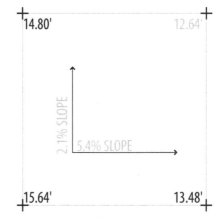

Figure 4.6A. Solution to slope problems

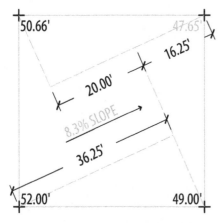

Figure 4.7A. Solution to slope problems

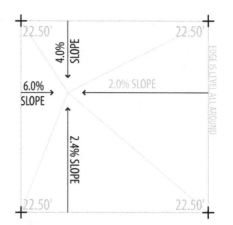

Figure 4.8A. Solution to slope problems

4.6

50:1 = 2%
10:1 = 10%
8:1 = 12.5%
3:1 = 33.3%

4.7

1.5% = 0°52'
3.5% = 2°0'
18% = 10°12'
35% = 19°17'

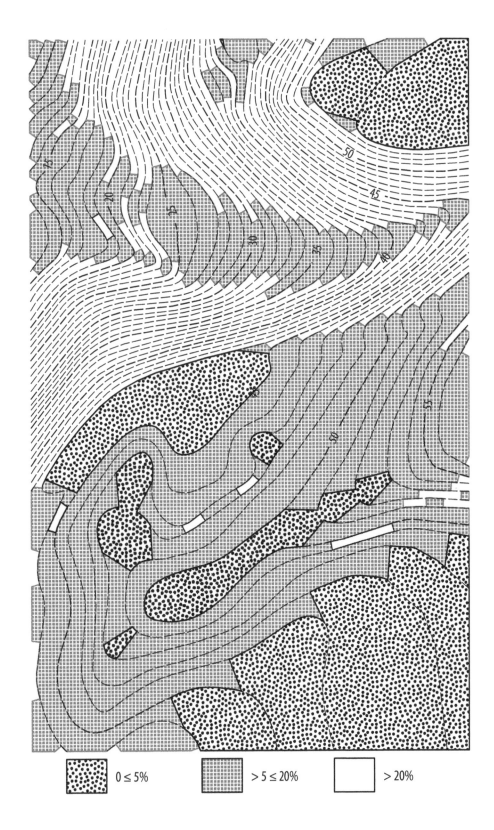

Figure 4.9A. Solution of slope analysis

0 ≤ 5%   > 5 ≤ 20%   > 20%

CHAPTER 5

Answers

5.1

The slope perpendicular to traffic on a sidewalk is called the **cross slope.** It is typically graded at **2** percent. ADA requirements specify this as a maximum percentage slope. Some municipalities will require that cross slopes fall within a range from 0.5 to 2 percent. Look up your local municipal code to see what design requirements are indicated.

5.2

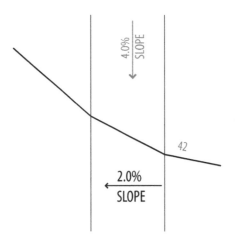

Figure 5.1A. Solution for cross-slope contour diagram

5.3

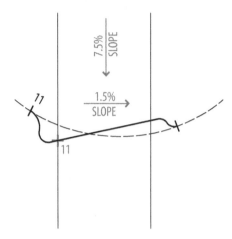

Figure 5.2A. Solution for cross-slope contour diagram

5.4

5.5

In order to avoid ponding on a roadway, a hazardous condition for drivers, crowns are designed to increase the speed of storm water runoff from the road surface. Crowns also help reinforce the separation between opposing lanes of traffic.

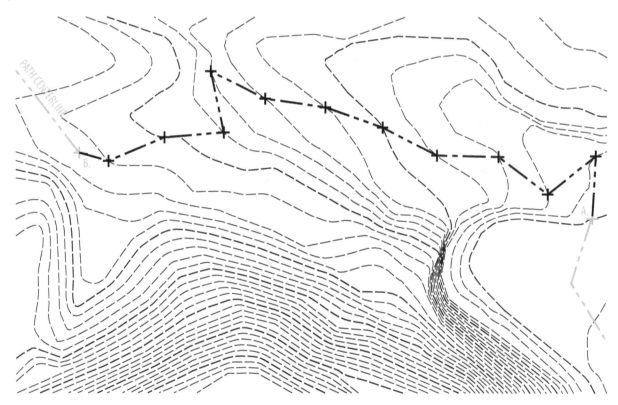

Figure 5.3A. Solution for path design

5.6

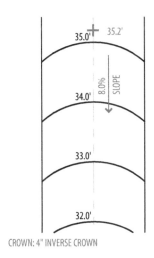

CROWN: 4" INVERSE CROWN

Figure 5.4A. Solution for road grading plan

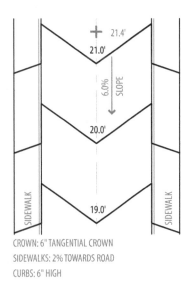

CROWN: 6" TANGENTIAL CROWN
SIDEWALKS: 2% TOWARDS ROAD
CURBS: 6" HIGH

Figure 5.6A. Solution for road grading plan

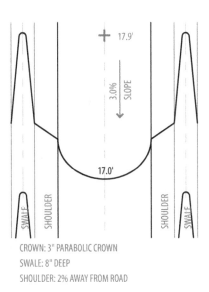

CROWN: 3" PARABOLIC CROWN
SWALE: 8" DEEP
SHOULDER: 2% AWAY FROM ROAD

Figure 5.5A. Solution for road grading plan

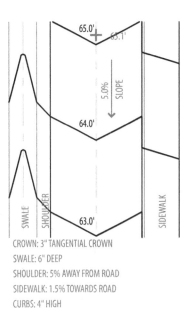

CROWN: 3" TANGENTIAL CROWN
SWALE: 6" DEEP
SHOULDER: 5% AWAY FROM ROAD
SIDEWALK: 1.5% TOWARDS ROAD
CURBS: 4" HIGH

Figure 5.7A. Solution for road grading plan

5.7

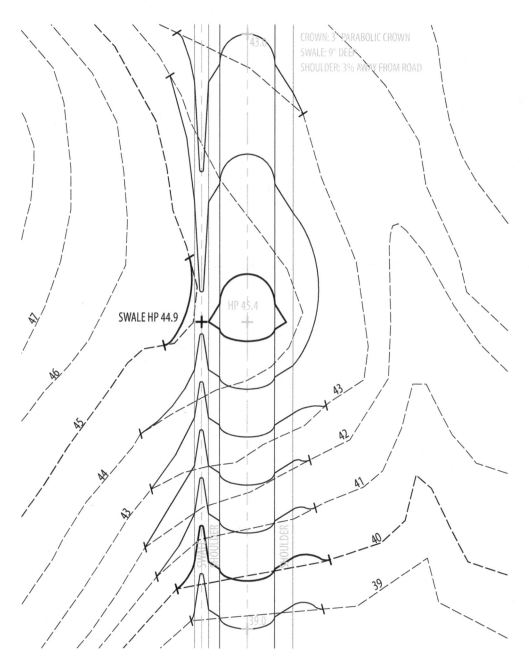

Figure 5.8A. Solution for road grading plan

5.8

Providing drainage away from a structure is referred to as providing **positive** drainage.

5.9

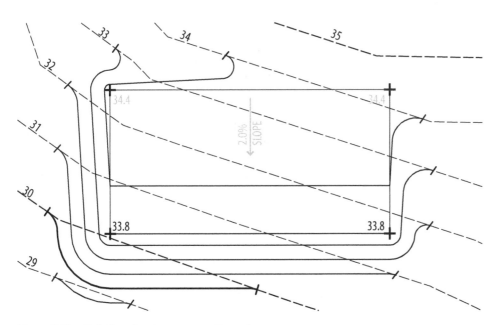

Figure 5.9A. Solution for terrace grading plan

5.10

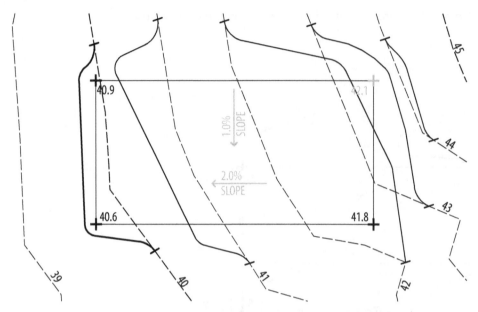

Figure 5.10A. Solution for terrace grading plan

CHAPTER 6
Answers

Please note that for all questions in this chapter, the answers provided are one of many that could meet the requirements laid out.

6.1

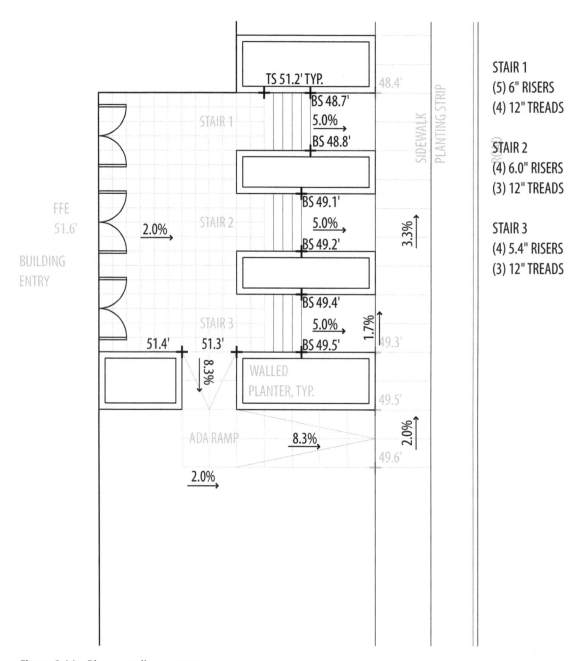

Figure 6.1A. Plaza grading answer

6.2

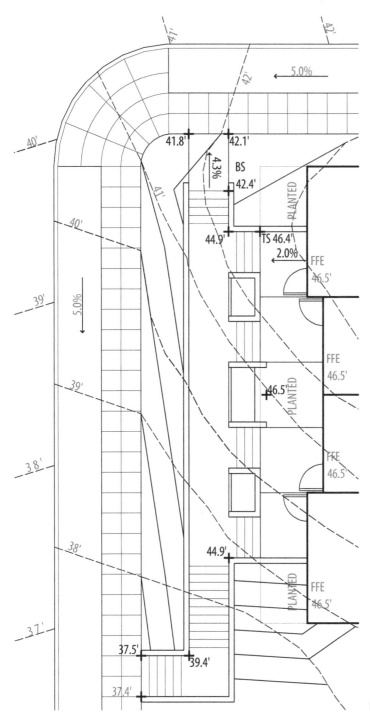

Figure 6.2A. Entry design answer

6.3

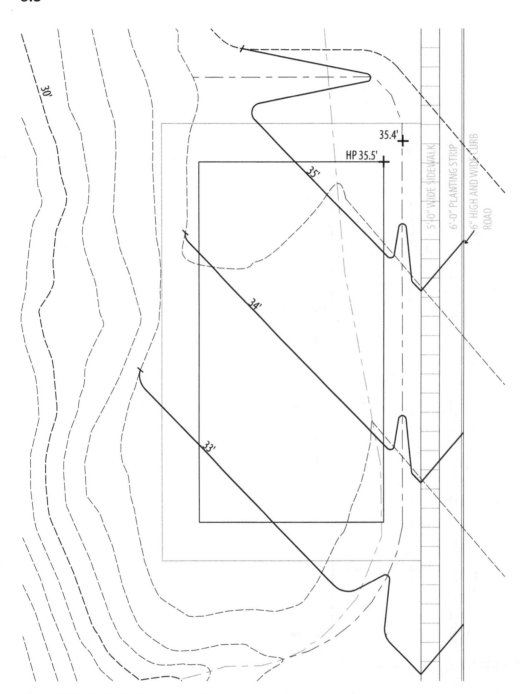

Figure 6.3A. Basketball court grading answer

6.4

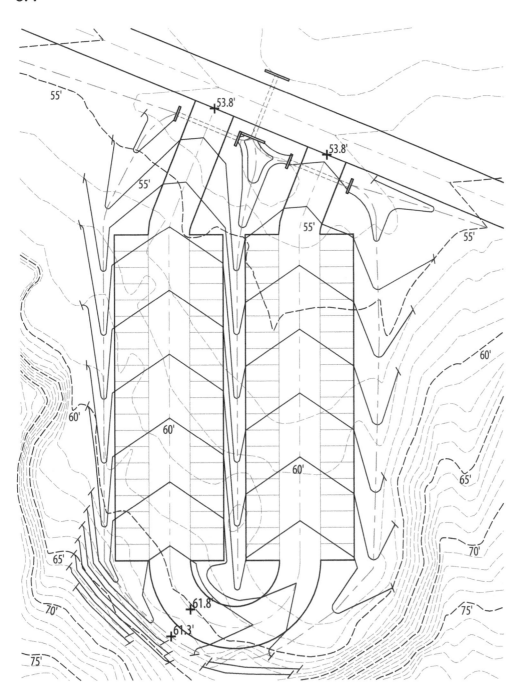

Figure 6.4A. Parking lot grading answer

6.5

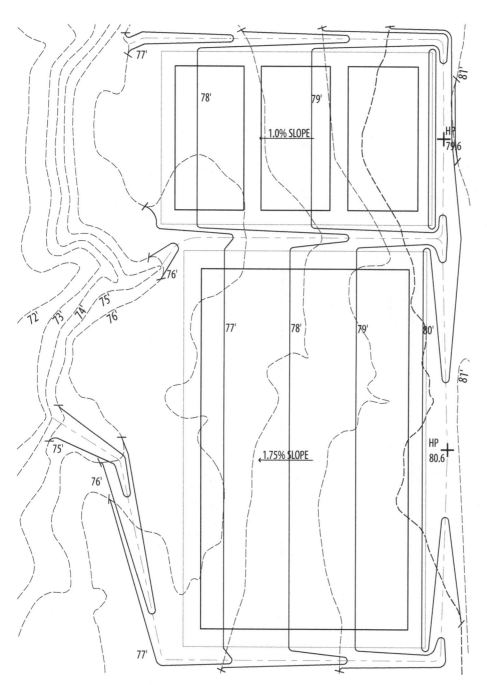

Figure 6.5A. Sports courts grading answer

6.6

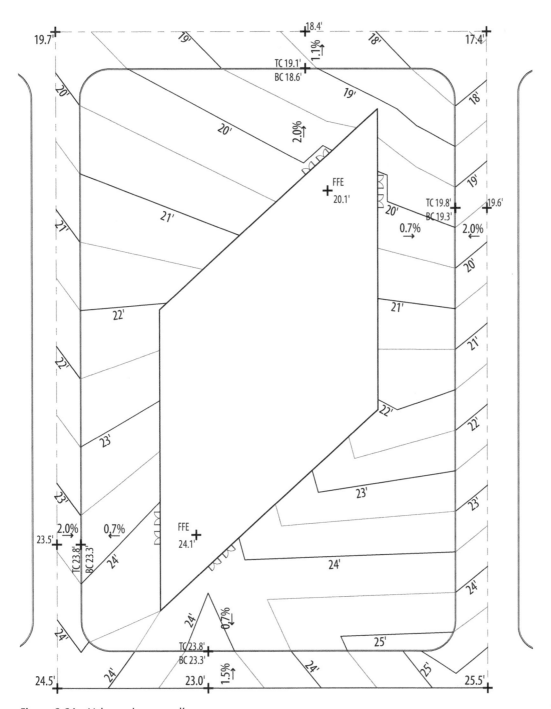

Figure 6.6A. Urban plaza grading answer

6.7

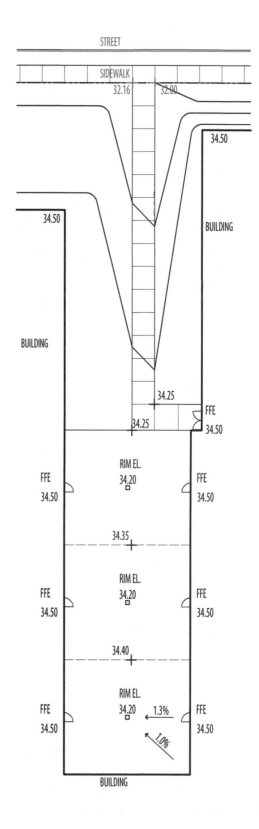

Figure 6.7A. Courtyard grading answer

6.8

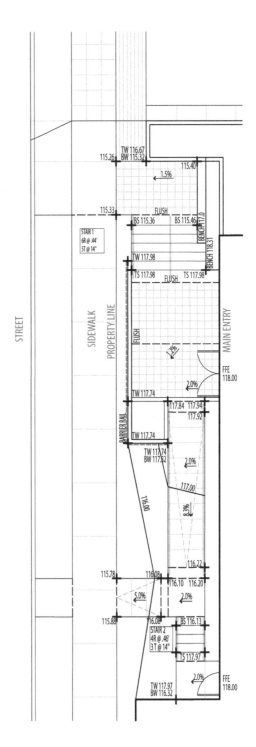

Figure 6.8A. Urban streetscape grading answer

CHAPTER 7

Answers

7.3

The shear strength of the exposed soils, maintenance considerations, and any additional stabilization technique to be employed all play a part in determining how steeply a slope can be graded without the construction of a retaining wall or other grade change device.

7.4

Loads applied to foundations and retaining walls in addition to the earth being retained are called **surcharges.**

7.5

Surcharges might take the form of slopes or other earthworks, structures, and vehicular traffic.

7.6

Depending on the moisture content of a soil, it can move between solid, semisolid, plastic, and liquid states. The shrinkage limit defines the boundary between solid and semisolid states. The plastic limit defines the boundary between semisolid and plastic states. The liquid limit defines the boundary between plastic and liquid states. Collectively, these limits are called the Atterberg limits.

7.7

A. 25% sand, 30% silt, 45% clay
B. 60% sand, 25% silt, 15% clay
C. 5% sand, 85% silt, 10% clay
D. 85% sand, 5% silt, 10% clay
E. 15% sand, 55% silt, 30% clay

Note that each set of percentages adds up to 100%.

7.8

A. Clay
B. Sandy clay
C. Silt loam
D. Sand
E. Clay loam

Figure 7.1A. Locations on USDA soil textural triangle

7.9

When examining a site and determining the potential for soil to support a structure, the soil's **bearing capacity** must be identified.

7.10

Amounts of internal friction and cohesion determine the shear strength of a soil.

7.11

Sand and gravel are cohesionless, whereas clay particles have a high amount of cohesion.

7.12

Slope failure occurs because of either (**increased/~~decreased~~**) stress or (~~increased~~/**decreased**) strength brought about by natural or human-induced activity.

7.13

(**True**/~~False~~) An increase in moisture content in soil can result in both decreased strength and increased stress of the soil.

7.14

Soil with high proportions of **clay** particles is prone to shrinking and swelling.

7.15

A soil that does not contain midrange-size stone is considered **gap-graded** and is a major component in the creation of **structural** soil.

7.16

In rocky alpine conditions, many plants grow in conditions that are essentially soil-less with little or no organic component.

7.17

Drainage, filtration, reinforcement, and separation are the four typical functions of geotextiles. Many geotextiles are designed to perform multiple functions in a single element.

7.18

By scarifying a landscape after compaction, permeability and aeration are improved, thus improving conditions for healthy vegetation.

7.19

CU soil is a mix of gap-graded angular crushed stone, clay loam soil, and hydrogel as a binding agent. CS soil is a mix of Stalite, a porous expanded slate rock, and sandy clay loam soil. Davis soil is a mix of lava rock and loam soil.

7.20

Structured soil volumes are designed to support paving or other surface structures, while creating a void that can be filled with ordinary topsoil.

7.21

1. Protection of existing vegetation
2. Topsoil removal and stockpiling
3. Erosion and sediment control
4. Clearing and demolition
5. Placement of grade stakes
6. Bulk excavation
7. Backfilling and fine grading
8. Finish surfacing

7.22

A **lift** is a uniform layer of fill installed in a specified thickness.

CHAPTER 8

Answers

8.1

Table 8.1A indicates that approximately 155 ft^3 of cut will be required. The following shows the conversion to yd^3:

$$155 \text{ ft}^3 \times 1 \text{ yd}^3/27 \text{ ft}^3 \approx 5.7 \text{ yd}^3$$

8.2

From Table 8.2A:

The amount of cut adjusted into cubic yards equals

$$573.9 \text{ ft}^3 \times 1/27 = 21.3 \text{ yd}^3$$

Table 8.1A. Station area data

Station	Area of Cut (ft²)	Area of Fill (ft²)	Average (ft²)	Distance (ft)	Volume (ft³)
0 + 00	0.0	10.0			
			42.7	50	2,135
0 + 50	0.0	75.3			
			97.1	50	4,855
1 + 00	−5.5	124.3			
			125.9	50	6,295
1 + 50	−24.7	157.6			
			156.3	50	7,815
2 + 00	−55.8	235.4			
			138.2	50	6,910
2 + 50	−78.6	175.3			
			41.3	50	2,065
3 + 00	−125.5	111.4			
			−83.9	50	−4,195
3 + 50	−178.9	25.3			
			−148.0	50	−7,400
4 + 00	−155.5	13.1			
			−125.1	50	−6,255
4 + 50	−107.7	0.0			
			−100.6	50	−5,030
5 + 00	−93.5	0.0			
			−80.6	50	−4,030
5 + 50	−67.7	0.0			
			−50.1	50	−2,505
6 + 00	−38.2	5.7			
			−16.3	50	−815
6 + 50	0.0	0.0			
Total					**−155**

The amount of fill adjusted into cubic yards equals

$$1{,}342.4 \text{ ft}^3 \times 1/27 = 49.7 \text{ yd}^3$$

Total amount of fill equals

$$49.7 - 21.3 = 28.4 \text{ yd}^3$$

The above measurements will vary slightly. depending on the method used to determine contour area.

Table 8.2A. Contour Area Measurements

Contour Number	Area of Cut (ft²)	Area of Fill (ft²)
40	230.7	
41	343.2	
42		975.2
43		338.8
44		**28.4**
Total	573.9	1,342.4

8.3

From Table 8.3A:

The amount of cut adjusted into cubic yards equals

$$3{,}073.3 \text{ ft}^3 \times 1/27 = 113.8 \text{ yd}^3$$

The amount of fill adjusted into cubic yards equals

$$2{,}534.3 \text{ ft}^3 \times 1/27 = 93.9 \text{ yd}^3$$

Total amount of cut equals

$$113.8 - 93.9 = 19.9 \text{ yd}^3$$

The above measurements will vary slightly depending on the method used to determine contour area.

Table 8.3A. Contour Area Measurements

Contour Number	Area of Cut (ft²)	Area of Fill (ft²)
39	131.6	107.8
40	250.7	33.7
41	295.5	25.9
42	549.1	
43	1,075.7	360.6
44	683.4	1,371.7
45	87.3	634.6
Total	**3,073.3**	**2,534.3**

8.4

From Figure 8.1A.

$$h_1 = 5.6 + 14.1 + 15.8 + 11.2 + 1.2 + 0.0$$
$$= 47.9$$
$$h_2 = 8.9 + 9.7 + 6.6 + 2.2 = 27.4$$

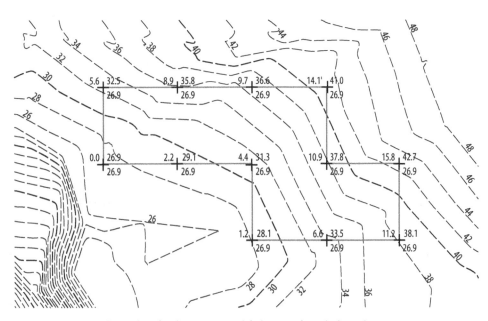

Figure 8.1A. Grading plan for basement with interpolated elevations

$h_3 = 4.4 + 10.9 = 15.3$
$h_4 = 0$
$A = 16 \text{ ft.} \times 16 \text{ ft.} = 256$
$V = A/4 \times (1\,h_1 + 2\,h_1 + 3\,h_1 + 4\,h_1)$
$\quad = 256/4 \times (1(47.9) + 2(27.4) + 3(15.3)$
$\quad\quad + 4(0))$
$\quad = 64\,(47.9 + 54.8 + 45.9)$
$\quad = 64 \times 148.6 = 9{,}510.4 \text{ ft}^3$
$9{,}510.4 \text{ ft}^3 \times 1 \text{ yd}^3/27 \text{ ft}^3 = 352.2 \text{ yd}^3$ of cut

8.5

The elevations provided on the plan in Figure 8.2A. are organized as follows: On the right side of the spot, the proposed elevation is on top, while the existing elevation is on the bottom. The difference between proposed and existing elevations is shown to the left of the spot elevation. All differences are positive, given that this problem was to design a terrace completely on fill.

With all perimeter spot elevations having a value of zero for the difference in elevation, the equation for calculating the total fill can be simplified in the following way:

$$V = A/4\,(4\,h_4) = A(h_4)$$

With A equal to 100 ft², the amount of fill required to construct the design is calculated as follows:

$$V_{\text{fill}} = A(h_4) = 100(53.8) = 5{,}380 \text{ ft}^3 \times 1/27 = 199.3 \text{ yd}^3$$

The volume of paving at the terrace should be subtracted from this total.

$$V_{\text{pav}} = \text{area of paving} \times \text{Depth of paving}$$

$$V_{\text{pav}} = 1{,}800 \text{ ft}^2 \times 0.33 \text{ ft.} = 594 \text{ ft}^3 \times 1/27 = 22.0 \text{ yd}^3 \text{ of paving material}$$

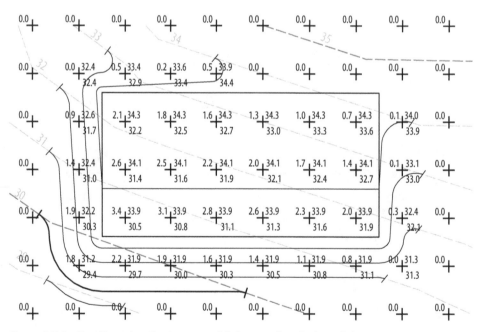

Figure 8.2A. Grading plan for terrace with interpolated elevations

The volume of existing topsoil is calculated by multiplying the area within the no cut–no fill boundary by the depth of topsoil.

$$V_{tse} = 3{,}642.9 \text{ ft}^2 \times 0.5 \text{ ft.} = 1{,}821.5 \text{ ft}^3 \times 1/27$$
$$= 67.5 \text{ yd}^3$$

The volume of proposed topsoil is calculated by multiplying the area within the no cut–no fill boundary, excluding the terrace area, by the depth of topsoil to be installed.

$$V_{tsp} = (3{,}642.9 - 1{,}800.0) \text{ ft}^2 \times 0.67 \text{ ft.}$$
$$= 1{,}234.7 \text{ ft}^3 \times 1/27 = 45.7 \text{ yd}^3$$

Subtracting the proposed topsoil placement from the existing topsoil shows that there is an amount of topsoil to be removed from the site as cut.

$$V_{tse} - V_{tsp} = 67.5 - 45.7 = 21.8 \text{ yd}^3$$ of topsoil to be removed from the site. This will require an additional amount of fill equivalent to the topsoil being removed to be added to the overall amount of fill.

The final amount of fill required can be calculated as follows:

$$V_{final} = V_{fill} - V_{pav} + V_{ts} = 199.3 - 22.0 + 21.8 = 199.1 \text{ yd}^3$$

CHAPTER 9

Answers

9.2

Loss of vegetation and organic litter and elimination of surface roughness and perviousness all play a part in increasing the rate and volume of runoff, often overwhelming existing drainage systems.

9.3

All of the following are considered impacts due to changes in the surface characteristics in a developed landscape:

- Increased flood potential due to increase in peak flow rates
- Decreased groundwater supply caused by reduced infiltration

- Increased soil erosion and sedimentation brought about by greater runoff volumes and velocities
- Increased petrochemical pollution from street and highway runoff
- Contamination of winter runoff by salt and sand in colder regions

9.4

(**True**/~~False~~) Higher velocities coupled with increased imperviousness may also result in reduced stream flow during extended dry periods.

This is caused by reduced infiltration of the faster-moving water through the system. Groundwater that would normally be recharged during wet periods and released slowly from soil during dry periods is lost as surface runoff.

9.5

Streams are widened by the increased volume and velocity of water; stream banks are often undercut, which disturbs vegetation and introduces additional sediment to the water; the additional sediment deposits itself in the streambed, which reduces the stream's capacity for water volume; this, in turn, raises flood elevations, as the water seeks room to move during times of high volume.

9.6

Pollutants accumulate and concentrate on the impervious surfaces. They are also quickly transported in rainstorms. Landscape development increases the number of sources of pollutants as well. Pesticides, herbicides, and fertilizers are also picked up into runoff and stimulate algae growth and deoxygenation of water. Nutrient and bacteria levels are also increased by the addition of leaves and animal droppings that would normally decompose in place in a more permeable environment but are readily picked up in the developed landscape and deposited into bodies of water.

9.7

Water quality, water volume, and rate of runoff are the three important factors that any integrated approach must address.

9.8

Degradation of water quality in streams is known to take place when the impervious surface coverage of a drainage area approaches approximately **20** percent.

9.9

Reducing the parking stall area where appropriate and studying the required parking demand ratio to better understand the project's actual need are two ways to manage the impact of developing parking lots through design. Both approaches may require obtaining a variance from a local municipality or even working with the municipality to develop a more appropriate standard.

9.10

Evapotranspiration, bioretainment, evaporation, detention, infiltration, capture, and treatment are all components being introduced into present-day storm water management systems to mitigate against the impacts of impervious surfaces in the built environment.

9.11

Sedimentation, filtration, thermal attenuation, adsorption, plant resistance, volatilization and phytoremediation are all types of treatment that can be programmed into storm water management systems.

9.12

Bioretainment is the storage of rainwater on the surfaces of a planting.

9.13

Soluble nutrients, metals, and organics are removed from storm water runoff through a process called **adsorption**.

CHAPTER 10

Answers

10.4

Box culverts have a rectangular cross section, while pipe culverts can have a circular, an elliptical, or an arch cross section.

10.5

The rule of thumb is that a single catch basin can be used for every 10,000 ft² of paved surface.

10.6

A drain inlet moves runoff directly into a drain pipe, while a catch basin has a sump that can trap sediment and debris before it enters the drain pipe.

10.7

BMPs may variously control, store, and/or treat storm water runoff.

10.8

Climate, soils, topography, proposed and existing land use, and surface cover should be studied as part of any analysis undertaken to determine appropriate BMPs for a site.

10.9

Increasing the length of flow between the inlet and outlet of a retention pond increases the time for pollutant and sediment settlement.

10.10

Ponds, basins, and paved areas can all be used to detain storm water on the landscape surface.

10.11

Dry wells, porous fill, oversized drainage structures, and cisterns can all be used to detain storm water in the landscape subsurface.

10.12

Rates of soil permeability, potential for groundwater pollution, and the possibility for a reduction in permeability rates over time are all limiting factors in deciding whether an infiltration structure is an appropriate BMP for a particular landscape.

10.13

Groundwater recharge, control of storm water runoff, enhancement of water quality, and promotion of biodiversity are all possible functions a constructed treatment wetland might perform.

10.14

A catchment area, a conveyance system, and a storage element are all required for a rainwater harvesting system to function.

10.15

The main difference between extensive and intensive green roofs is their depth of soil and, as a result, the types of plants they are able to support. Extensive green roofs have a depth of 2 to 6 in. of soil. Intensive roofs will have greater than 6 in. of soil depth.

10.16

This layer of material provides aeration to the growing medium so that it does not become waterlogged. Some systems are also designed to hold a small amount of moisture to mitigate some of the impacts of drought.

10.17

Green roof systems that have been designed to provide habitat are called **brown roofs**.

10.18

Where bioretainment depends on the plant structure for runoff reduction, bioretention combines the bioretainment benefits of a layered planting of trees, shrubs, and groundcovers with a designed soil that promotes infiltration.

10.19

One hundred percent of occupants' water use must come from captured precipitation or closed-loop water systems that account for downstream ecosystem impacts and are appropriately purified without the use of chemicals.

One hundred percent of storm water and building water discharge must be managed on-site to feed the project's internal water demands or released onto adjacent sites for management through acceptable natural time-scale surface flow, groundwater recharge, agricultural use, or adjacent building need.

CHAPTER 11

Answers

11.3

Erosion and sedimentation cycle through initial detachment of soil particles, transportation of those soil particles, and deposition. Those new soil deposits may at a later date be picked up and moved, repeating the cycle.

11.4

Rainfall impact, flowing water, freezing and thawing, and wind can all cause soil detachment.

11.5

Soil type, vegetative cover (or lack thereof), topography, and climate must all be considered when trying to understand a soil's erosion potential.

11.6

Silt and fine sands are the most erodible; well-sorted gravel and gravel-sand mixes are the least erodible; and clay falls in the middle.

11.7

With respect to vegetative cover, the goal of most site development projects should **be to retain as much existing vegetation as possible, especially in areas of steep slopes, stream banks, drainageways, or areas of poor soils.**

11.8

Working with existing topography, restricting the area of disturbance, developing compactly, managing site construction, and preserving existing vegetation are all important principles in minimizing site disturbance.

11.9

Runoff control measures can generally be grouped as diversions, waterway stabilization, slope protection structures, grade control structures, and outlet protection.

11.10

Soil bioengineering is the use of live, woody, and herbaceous plants to stabilize or protect stream banks, shorelines, drainageways, and upland slopes. It combines biological and ecological concepts with engineering principles to prevent or minimize slope failure and erosion. Vegetation may be used alone or in combination with structural elements, such as rock, wood, concrete, or geotextiles.

CHAPTER 12

Answers

12.1

The Rational method assumes three things:

- The rainfall intensity of a storm is uniform for the duration of the storm.
- The duration of the storm and the time of concentration are equal.
- Precipitation falls on the entire drainage area for the duration of the storm.

12.2

When values of the runoff coefficient, C, get closer to 1, they are becoming (~~more~~/**less**) pervious.

12.3

Rainfall intensity is a measurement that represents both the design storm frequency and the time of concentration of the drainage area.

12.4

(~~True~~/**False**) A 20-year design storm will happen once every 20 years in a particular location.

This statement is false because the storm frequency is a representation of the probability that a storm will happen in that amount of time. In some 20-year periods, a storm of the intensity of a 20-year design storm may happen more than once; in other 20-year periods, it may not happen at all.

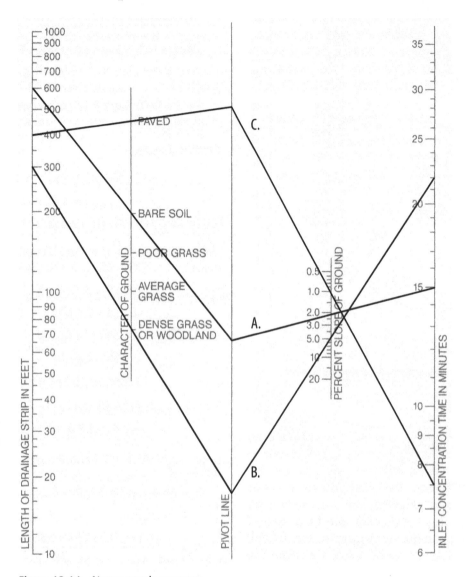

Figure 12.1A. Nomograph answers

12.5

The time of concentration is the time it takes for water to flow from the most hydraulically remote part of the drainage area to the section under consideration. This may not represent the longest distance, since rate of flow varies, depending on slope, surface, and channel characteristics.

12.6

From Figure 12.1A.

A. Time of concentration = 15 minutes.

B. Time of concentration = 22 minutes.

C. Time of concentration = 7.5 minutes.

12.7

Given from problem description:

$$A_{roadway} = 1.2 \text{ ac}$$
$$A_{rooftop} = 1 \text{ ac}$$
$$A_{woodland} = 2.4 \text{ ac}$$
$$A_{lawn} = 4.2 \text{ ac}$$
$$i = 2.5 \text{ iph}$$

Obtained from the table:

$$C_{roadway} = 0.70$$
$$C_{rooftop} = 0.95$$
$$C_{woodland} = 0.50$$
$$C_{lawn} = 0.10$$
$$A_{total} = A_{roadway} + A_{rooftop} + A_{woodland} + A_{lawn}$$
$$A_{total} = 1.2 + 1 + 2.4 + 4.2 = 8.8$$
$$C_{ave.} = [(A_{roadway} \times C_{roadway}) + (A_{rooftop} \times C_{rooftop}) + (A_{woodland} \times C_{woodland}) + (A_{lawn} \times C_{lawn})]/A_{total}$$
$$C_{ave.} = [(1.2 \times 0.70) + (1 \times 0.95) + (2.4 \times 0.50) + (4.2 \times 0.10)]/8.8$$
$$C_{ave.} = [(0.84) + (0.95) + (1.2) + (0.42)]/8.8$$
$$C_{ave.} = 3.41/8.8$$
$$C_{ave.} = 0.3875 \approx 0.39$$
$$q_{total} = 0.39 \times 2.5 \times 8.8$$
$$q_{total} \approx 8.6 \text{ ft}^3/\text{s}$$

12.8

The paved surface results in 9.5 minutes. The poor grass surface results in 10 minutes. The dense grass surface results in 11 minutes.

To find the contribution of the stream flow, H must be calculated by multiplying the length of flow by the slope: $H = (1,600 \times 0.01) = 16 \text{ ft}$.

Using the calculated H, the stream flow results in a 13-minute time of concentration.

Adding the results of the four segments produces the total time of concentration: $9.5 + 10 + 11 + 13 = 43.5$.

12.9

When using the Rational method, the duration of the storm is the same as the time of concentration. Using the 43.5-minute TC calculated in question 12.8 and the 50-year frequency given in the problem description, the rainfall intensity, i, is equal to approximately 3.2 iph. Remember that the data in the figure are logarithmic.

Given from the problem description:

$$A_{cemetery} = 14 \text{ ac}$$
$$A_{railroad} = 6 \text{ ac}$$
$$A_{industry} = 20 \text{ ac}$$
$$A_{unimproved} = 26 \text{ ac}$$

Obtained from the table provided in question 12.7:

$$C_{cemetery} = 0.25$$
$$C_{railroad} = 0.35$$
$$C_{industry} = 0.80$$
$$C_{unimproved} = 0.30$$
$$A_{total} = A_{cemetery} + A_{railroad} + A_{industry} + A_{unimproved}$$
$$A_{total} = 14 + 6 + 20 + 26 = 66$$
$$C_{ave.} = [(A_{cemetery} \times C_{cemetery}) + (A_{railroad} \times C_{railroad}) + (A_{industry} \times C_{industry}) + (A_{unimproved} \times C_{unimproved})]/A_{total}$$
$$C_{ave.} = [(14 \times 0.25) + (6 \times 0.35) + (20 \times 0.80) + (26 \times 0.30)/66$$
$$C_{ave.} = [3.5 + 2.1 + 16 + 7.8]/66$$
$$C_{ave.} = 29.4/66$$
$$C_{ave.} = 0.445 \approx 0.45$$
$$q_{total} = 0.45 \times 3.2 \times 66 = 95.04$$
$$q_{total} \approx 95.0 \text{ ft}^3/\text{s}$$

12.10

Peak runoff rates can be used to calculate the size of culverts, waterways, and pipes.

12.11

Inflow and outflow volumes must be calculated to understand how to size structures that provide storage, such as storage ponds and reservoirs.

12.12

For a type A hydrograph, the duration of the storm event is **equal to** the time of concentration.

For a type B hydrograph, the duration of the storm event is **greater than** the time of concentration.

For a type C hydrograph, the duration of the storm event is **less than** the time of concentration.

12.13

Given from the problem description:

$$A = 19 \text{ ac}$$

Obtained from chapter figures:

    $C = 0.30$ (from the table in question 11.7)

    $C_A = 1.25$ (from the table in question 11.13)

    $i \approx 6.5$ iph (from a 20-minute time of concentration using the figure in question 11.6)

    $q = C\ C_A\ i\ A = 0.30 \times 1.25 \times 6.5 \times 19$
    $= 46.31$

    $q_{total} \approx 46.3 \text{ ft}^3/\text{s}$

12.14

See the figures below for hydrographs. Calculations for obtaining the hydrographs are found below.

10-minute hydrograph

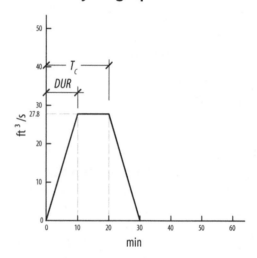

Figure 12.2A. 10-minute hydrograph

Given from the problem description:

$$A = 19 \text{ ac}$$

Obtained from chapter figures:

    $C = 0.30$ (from the table in question 11.7)

    $C_A = 1.25$ (from the table in question 11.13)

    $i \approx 7.8$ iph (from a 10-minute time of concentration using the figure in question 11.6)

    $q_{max} = C\ C_A\ i\ A\ (DUR/T_C) = 0.30 \times 1.25$
    $\times 7.8 \times 19 \times (10/20) = 27.7875$

    $q_{max} \approx 27.8 \text{ ft}^3/\text{s}$

20-minute hydrograph

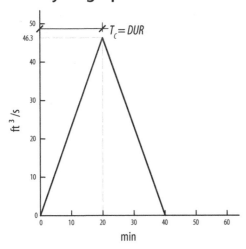

Figure 12.3A. 20-minute hydrograph

See answer, q, in question 12.13 for q_p.

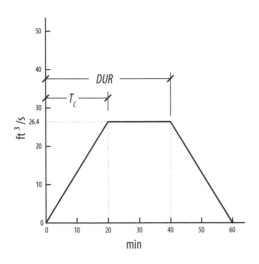

Figure 12.4A. 40-minute hydrograph

40-minute hydrograph

Given from the problem description:

$$A = 19 \text{ ac}$$

Obtained from chapter figures:
   $C = 0.30$ (from the table in question 12.7)
   $C_A = 1.25$ (from the table in question 12.13)
   $i \approx 3.7$ iph (from a 10-minute time of concentration using the figure in question 12.6)
   $q_{max} = C\,C_A\,i\,A = 0.30 \times 1.25 \times 3.7 \times 19$
      $= 26.3625$
   $q_{max} \approx 26.4 \text{ ft}^3/\text{s}$

12.15

Given from the problem description:

$$A = 19 \text{ ac}$$

Obtained from chapter figures:
   $C = 0.70$ (from the table in question 12.7)
   $C_A = 1.25$ (from the table in question 12.13)
   $i \approx 7.8$ iph (from a 10-minute time of concentration using the figure in question 12.6)
   $q = C\,C_A\,i\,A = 0.70 \times 1.25 \times 7.8 \times 19 = 129.68$
   $q_{total} \approx 129.7 \text{ ft}^3/\text{s}$

12.16

As seen in Table 12.1A., the duration of the critical storm will be 15 minutes, and the maximum storage volume required will be $\approx 58,600 \text{ ft}^3$.

Table 12.1A. Solution

Storm Freq. (yr)	Duration (min.)	Intensity (iph)	C_A	Maximum Inflow (ft³/s)	Maximum Outflow (ft³/s)	Inflow Volume (ft³)	Outflow Storage (ft³)	Required Storage (ft³)
100	10	7.8	1.25	129.7	46.3	77,820	27,780	50,040
100	15	6.7	1.25	111.4	46.3	100,249	41,670	58,579
100	20	5.6	1.25	93.1	46.3	111,720	55,560	56,160

Answers

13.1

In using the NRCS method for estimating runoff, inches of precipitation are translated into inches of runoff using a **curve number,** which is based on soils, land use, impervious areas, interception by vegetation and structures, and temporary surface storage.

13.2

Flow is separated into sheet flow, shallow concentrated flow, and open channel flow in the TR-55 manual.

CHAPTER 14

Answers

14.1

The scale and intensity of development, the amount and location of paved and unpaved surfaces, the proposed uses, ecological impacts, and aesthetic concerns are all factors that must be taken into consideration when designing a drainage system. Soil erodibility, extent and steepness of slopes, and expected rainfall intensities are also factors to be considered.

14.2

Collection, conduction, and disposal are the three main functions of any storm drainage system.

14.3

(**True**/~~False~~) It is illegal to increase or concentrate flow across landscape outside of your project's property line.

14.4

Swales and pipes generally (**increase**/~~decrease~~) in size as they progress toward the outlet point. This is a result of the increased volume of water collected into the system closer to the outlet point.

14.5

Velocity could be reduced by a change in slope from steep to flat. It can also be reduced by enlarging the cross section without also increasing the slope, which causes the water to slow down due to increased surface area contact of the flow. An increase in the frictional resistance of the surface caused by tall vegetation being placed downslope from a stand of short vegetation can also reduce the velocity.

14.6

Reducing the velocity of water flowing in a swale results in **siltation.** The slowed water has reduced ability to carry the sediment, so it falls out of suspension and deposits. Over time this can change the character of the designed feature and affect the swale's ability to function properly.

14.7

The type and condition of vegetation, the erodibility of the soil, and the slope of the swale all factor into the permissible maximum design velocity of a swale.

14.8

$$q = AV$$

$V = q/A = 49/14 = 3.5 =$ velocity of flow in the parabolic swale described in question 13.11.

14.9

$$V = (1.486/n)\ R^{2/3}\ S^{1/2}$$
$$R^{2/3} = (4 \times 0.05)/(1.486 \times 0.035^{1/2})$$
$$R^{2/3} = 0.2/0.278$$
$$R^{2/3} = 0.72$$
$$R = 0.72^{1.5}$$
$$R = 0.61$$

14.10

$$q = pr^2(1.486/n)(r/2)^{0.67}\ S^{0.50}$$
$$r^2 \times r^{0.67} = (q \times n \times 2^{0.67})/(p \times 1.486 \times S^{0.50})$$
$$r^{2.67} = (4.5 \times 0.015 \times 1.59)/(p \times 1.486 \times 0.012^{0.50})$$
$$r^{2.67} = (4.5 \times 0.015 \times 1.59)/(p \times 1.486 \times 0.110)$$
$$r^{2.67} = 0.107/0.514$$
$$r^{2.67} = 0.208$$
$$r = 0.208^{1/2.67}$$
$$r = 0.56$$

14.11

Laterals

$$360\,\text{ft} \times 1{,}200\,\text{ft} = 432{,}000\,\text{ft}^2 \times 1\,\text{ac}/43{,}560\,\text{ft}^2$$
$$\approx 9.9\ \text{ac}$$

At a 0.2% slope, the laterals need to be sized to a diameter of 5 in. to handle the drainage.

Main line

$$(180 \text{ ft} \times 2,400 \text{ ft}) + (6 \times 432,000 \text{ ft}^2)$$
$$= 7 \times 432,000 \text{ ft}^2 = 3,024,000 \text{ ft}^2$$
$$\times 1 \text{ ac}/43,560 \text{ ft}^2 \approx 69.4 \text{ ac}$$

At a 0.6 percent slope, the main line needs to be sized to a diameter of 8 in. to handle the drainage.

14.12

4.3 in. / 12 in. = 0.358 feet of rainfall

$0.358 \times 0.80 = 0.2864$ cubic feet of water can be captured from every square foot of roof surface

4500/0.2864 = 15,713 s.f. of roof area required.

14.13

0.75 in. of irrigation water needed × 4 weeks = 3 in.

3 in. / 12 in. = 0.25 cubic feet of water needed per square foot of planted area to cover a four-week drought.

500/ 0.25 = 2,000 cubic feet of cistern volume required

CHAPTER **15**

Answers

15.2

The purpose of a layout plan is to establish **horizontal** position, orientation, and extent of all proposed construction elements. By contrast, **vertical** position is established by the grading plan.

15.3

The majority of site dimensions are **semifixed** dimensions. These dimensions generally locate site features that might include walks, plazas, parking areas, walls, lights and signs, and other landscape elements.

15.4

Movable furnishings are generally not located, and neither are plantings.

15.5

The measurement 34.6 ft implies a preciseness of **±0.05 ft.**

15.6

The starting point of a layout plan is called the point of beginning.

15.7

The problem with the expression is that when dimensions are less than a foot, feet are not indicated.

15.8

The perpendicular offset system is the most appropriate method of layout when site elements

137

are located orthogonal to property lines or proposed buildings.

15.9

The baseline layout system.

15.10

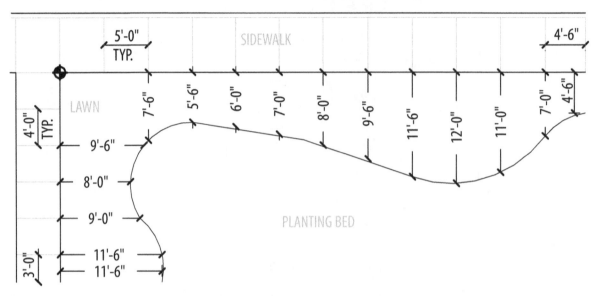

Figure 15.1A. One possible solution for the layout

15.11

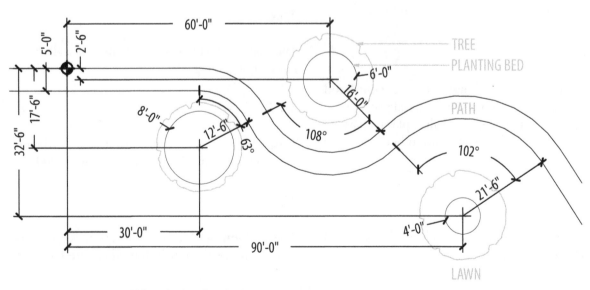

Figure 15.2A. One possible solution for the layout

15.12

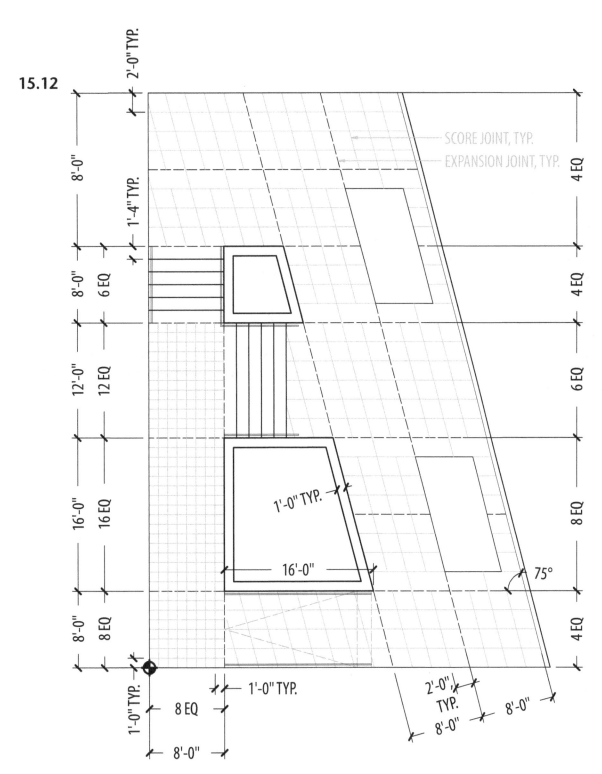

Figure 15.3A. One possible solution for the layout

15.13

The first advantage is the elimination of the clutter of dimension lines, as the coordinate information is typically held in a table. The second advantage is that, as opposed to a string of dimensions in which errors can become cumulative, coordinates are independent measurements, thus eliminating cumulative error problems. One drawback is that in an iterative design process, updating coordinate information can be time-consuming and changes can be difficult to track.

15.14

A bearing is an **acute** angle, measured off of the **north/south** axis or meridian.

15.15

A center point, a radius, and an internal angle are all needed to define an arc.

15.16

The accuracy of a GPS receiver is dependent on the equipment being used, line of sight to the sky, integration with other systems, and postprocessing of information.

15.17

At least four and preferably six satellites are required to triangulate a GPS receiver's signal and provide latitude, longitude, and elevation.

CHAPTER **16**
Answers

16.2

Tangents and curves are the basic geometric components of a horizontal road alignment.

16.3

Simple curves, compound curves, and reverse curves are the basic curve configurations most commonly used in horizontal road alignment. Broken-back curves are typically not recommended and should be replaced by a single larger curve.

16.4

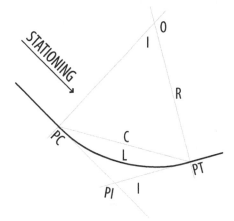

Figure 16.1A. Labeled curve plan answer

16.5

T = tangent distance: the distance from the PI to either the PC or the PT. These distances are always equal for simple circular curves.

I = included angle: the central angle of the curve, which is equal to the deflection angle between the tangents.

R = radius: the radius of the curve.

L = length of curve: the length of the arc from PC to PT.

C = chord: the distance from PC to PT measured along a straight line.

O = center of curve: the point about which the included angle I is turned.

PI = point of intersection: the point at which the two tangent lines intersect.

PT = point of tangency: the point that marks the end of the curve at which the road alignment returns to a tangent line in the direction of stationing.

PC = point of curvature: the point that marks the beginning of the curve at which the road alignment diverges from the tangent line in the direction of stationing.

16.6

(**True**/~~False~~) For simple horizontal curves, the distance from PC to PI and from PI to PT is always equal.

16.7

(~~True~~/**False**) A new road stationing system is usually started for each new road at the curb line of an existing road. The new stationing is always started at the centerline of an existing road.

16.8

The rise of the outer edge of pavement relative to the inner edge at a curve in the highway, expressed in feet per foot, intended to overcome the tendency of speeding vehicles to overturn when rounding a curve, is called **superelevation.** Its value should not exceed **0.083 ft/ft** under most conditions, or **0.125 ft/ft** if snow and ice are a local problem.

16.9

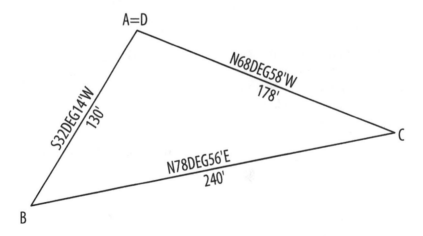

Figure 16.2A. Bearings

16.10

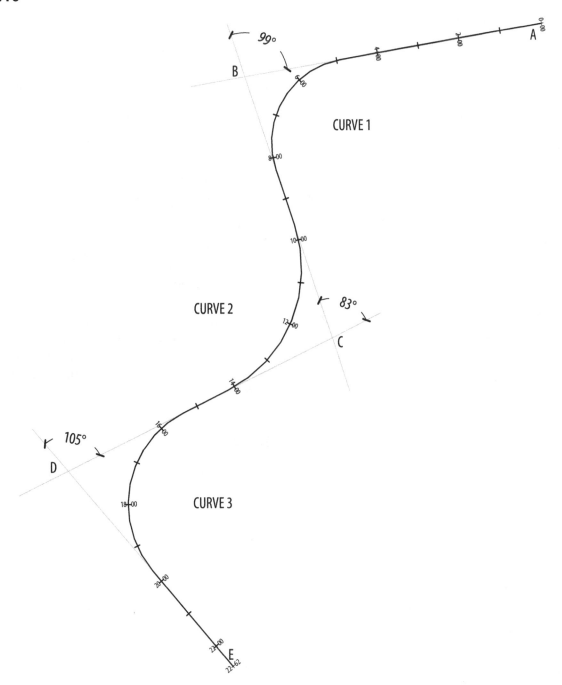

Figure 16.3A. Bearings, curves, and stationing

Curve 1

$L = ((2\pi R) \times I)/360$

$L = ((2\pi 195) \times 95°51')/360$

$L = 326.22'$

$T = R\tan(I/2)$

$T = 195 \times \tan(95°51'/2)$

$T = 216.00'$

$C = 2R\sin(I/2)$

$C = 2 \times 195 \times \sin(95°51'/2)$

$C = 289.48'$

Curve 2

$R = T/(\tan(I/2))$

$R = 276.49/(\tan(82°33'/2))$

$R = 315.00'$

$L = ((2\pi R) \times I)/360$

$L = ((2\pi 315) \times 82°33')/360$

$L = 453.84'$

$C = 2R\sin(I/2)$

$C = 2 \times 195 \times \sin(95°51'/2)$

$C = 415.59'$

Curve 3

$L = ((2\pi R) \times I)/360$

$L = ((2\pi 235) \times 104°39')/360$

$L = 429.22'$

$T = R\tan(I/2)$

$T = 235 \times \tan(104°39'/2)$

$T = 304.33'$

$C = 2R\sin(I/2)$

$C = 2 \times 235 \times \sin(104°39'/2)$

$C = 372.00'$

16.11

Figure 16.4A shows one possible solution to the requirements laid out in question 16.11. A radius of 450 ft was used to design the curve. The figure is for reference and is not shown at full scale.

$L = ((2\pi R) \times I)/360$

$L = ((2\pi 450) \times 42°18')/360$

$L = 332.22'$

$T = R\tan(I/2)$

$T = 450 \times \tan(42°18'/2)$

$T = 174.09'$

$C = 2R\sin(I/2)$

$C = 2 \times 450 \times \sin(42°18'/2)$

$C = 324.73'$

An argument could be made for creating a road alignment that fits better into the landscape. If the alignment were moved toward the upper left, the road could navigate around the ridge that it currently cuts across.

16.12

$R = 5{,}729.578/D = 5{,}729.578/20$

$R = 286.48'$

$L = ((2\pi R) \times I)/360$

$L = ((2\pi\, 286.48') \times 45°00')/360$

$L = 225'$

$T = R\tan(I/2)$

$T = 286.48' \times \tan(45°00'/2)$

$T = 118.66'$

$C = 2R\sin(I/2)$

$C = 2 \times 286.48' \times \sin(45°00'/2)$

$C = 219.26'$

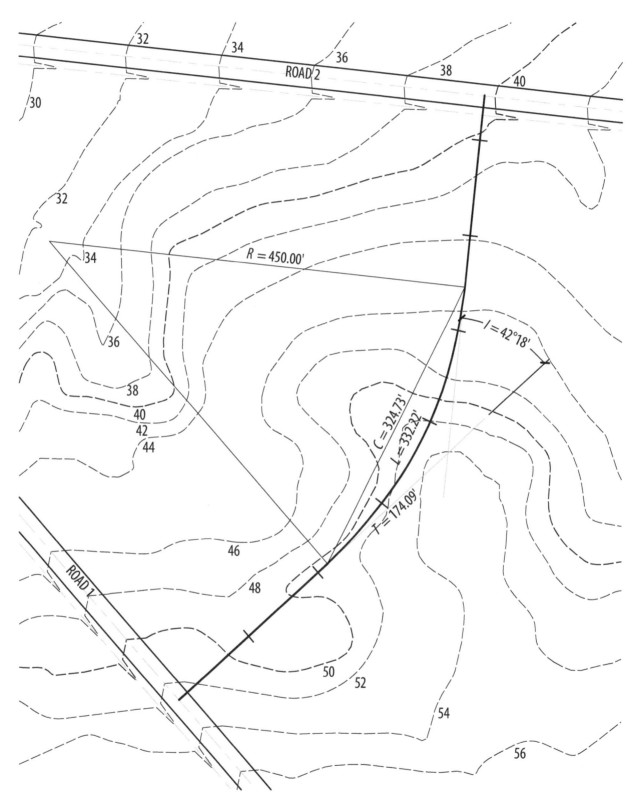

Figure 16.4A. Bearings, curve, and stationing

16.13

$S = 0.067 \, (V^2/R)$.

$S = 0.067 \, (40^2/375) = 0.067 \times 4.2667$
$= 0.29 \, \text{ft/ft of width}$

This exceeds the recommended maximum, so the preferred maximum value of 0.125 ft./ft., or 1.5 in./ft., is the proposed superelevation.

$$(0.25 + 1.5) \times 160 = 280 \, \text{ft. of runoff distance}$$

CHAPTER **17**

Answers

17.2

Tangent lines for horizontal curves are **direction** lines in the horizontal plane, whereas tangent lines for vertical curves are **slope** lines in the vertical plane.

17.3

Peak curves, in which the entering tangent gradient is positive and the exiting tangent gradient is negative in the direction of stationing. Sag curves, in which the entering gradient is negative and the exiting gradient is positive. Intermediate peak or sag curves occur when the change in slope occurs in the same direction—that is, both values either positive or negative.

17.4

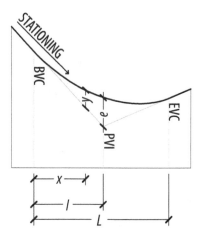

Figure 17.1A. Labeled curve plan

17.5

e = tangent offset at PVI

l = one-half length of curve

L = length of curve

x = horizontal distance from BVC (or EVC) to point on the curve

y = tangent offset distance

PVI = point of vertical intersection

BVC = beginning of vertical curve

EVC = end of vertical curve

17.6

(**True**/~~False~~) The formulas and calculations used to determine road curves can also be used to lay out any other curved features, including walls, fences, and pathways.

17.7

A vertical curve with a −2.1 percent slope coming into the PVI and a −5.7 percent slope leaving the PVI has an algebraic difference of **3.6** percent.

17.8

Tangent 1 is sloping at −1.7%.
    Tangent 2 is sloping at +6.1%.

$A = 7.8$

$L = 350'$

$l = 175'$

$e = ((175)^2/200 \times 2 \times 175) \times 7.8$
    $= 30{,}625/70{,}000 \times 7.8 = 3.4$

$d = Lg_1/(g_1 - g_2) = 350 \times -0.017/(-0.017 - 0.061) = 76.3$

$y = e(x/l)^2 = 3.4((76.3)/175)^2 = 0.65$

See Table 17.1A for all vertical curve elevations.

Table 17.1A. Vertical curve data

Station	Point	Tangent Elevation	Tangent Offset	Curve Elevation
0 + 50	BVC	52.2	0	52.2
1 + 00		51.4	0.28	51.68
1 + 26.3	LP	50.9	0.65	51.55
1 + 50		50.5	1.11	51.61
2 + 00		49.7	2.50	52.20
2 + 25	PVI	49.2	3.41	52.61
2 + 50		50.8	2.50	53.30
3 + 00		53.8	1.11	54.91
3 + 50		56.9	0.28	57.18
4 + 00	EVC	59.9	0	59.9

17.9

See Figure 17.2A.

Curve 1

Tangent 1 is sloping at −3.0 percent.
    Tangent 2 is sloping at +1.4 percent.
    $A = 4.4$
    $L = 150'$
    $l = 75'$
    $e = ((75)^2/200 \times 2 \times 75) \times 4.4 = 5{,}625/30{,}000 \times 4.4 = 0.83$
    $d = Lg_1/(g_1 - g_2) = 150 \times -0.030/(-0.030 - 0.014) = 102.3$
    $y = e(x/l)^2 = 0.83((102.27 - 75)/75)^2 = 0.34$

Elevation of the low point is found by measuring the slope back from the EVC along tangent 2 in this case, because the value of d was greater than the value of l. If the elevation at EVC 1 is 50.2, measuring toward the PVI 47.73 $(102.27 - l)$, at a slope of 1.4 percent, puts the elevation along the tangent at the LP at 49.53. Add y (0.34) to that and the elevation of the LP of the curve is 49.87.

Curve 2

Tangent 2 is sloping at $+1.4$ percent.
 Tangent 3 is sloping at -3.3 percent.

$$A = 4.7$$
$$L = 200'$$
$$l = 100'$$
$$e = ((100)^2/200 \times 2 \times 100) \times 4.7$$
$$= 10{,}000/40{,}000 \times 4.7 = 1.18$$
$$d = Lg_1/(g_1 - g_2) = 200 \times 0.014/(0.014$$
$$- (-0.033)) = 59.6$$
$$y = e(x/l)^2 = 1.18(59.6/100)^2 = 0.42$$

Elevation of the high point is found by measuring the slope from the BVC along tangent 2. If the elevation at BVC 1 is 51.2, measuring toward the PVI 59.6, at a slope of 1.4 percent, puts the elevation along the tangent at the LP at 52.03. Subtract y (0.42) from that and the elevation of the HP of the curve is 51.6.

17.10

Looking at the grading in Figure 17.3A, there are some observations worth noting regarding how the vertical alignment of the road might be changed to better fit the landscape. Setting a higher PVI for curve 2 would allow the ridgeline to be accommodated, without being quite so sharply cut off. At

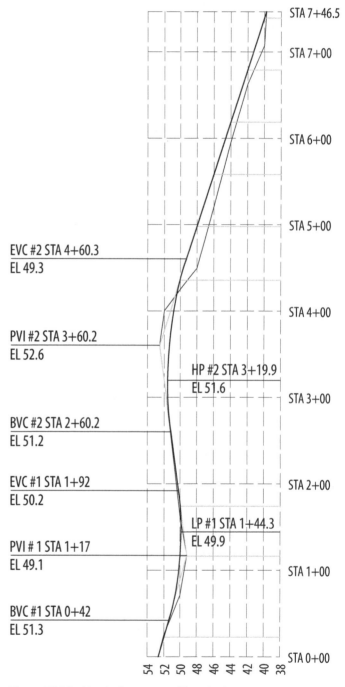

Figure 17.2A. Vertical curve profile solution

the beginning and end of the alignment, grading should take into account more directly the existing grades in the existing roads. The intersection at road 2 especially causes a disruption of the road grading that would be very noticeable. Figure 17.3A is for reference and is not drawn to full sale.

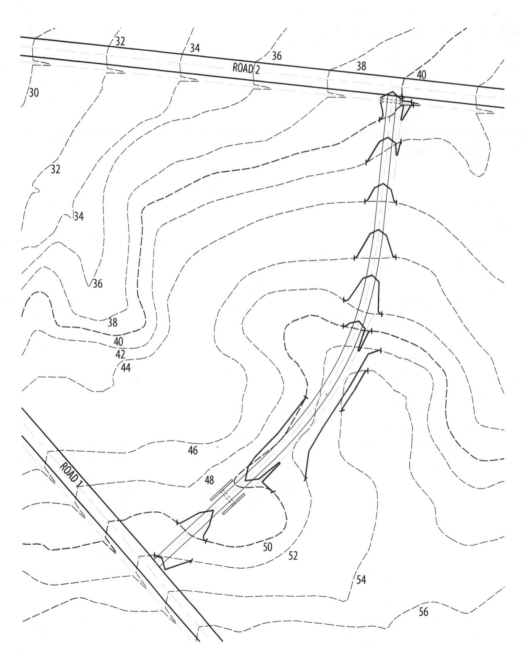

Figure 17.3A. Grading plan solution